About Island Press

Island Press is the only nonprofit organization in the United States whose principal purpose is the publication of books on environmental issues and natural resource management. We provide solutions-oriented information to professionals, public officials, business and community leaders, and concerned citizens who are shaping responses to environmental problems.

In 1994, Island Press celebrated its tenth anniversary as the leading provider of timely and practical books that take a multidisciplinary approach to critical environmental concerns. Our growing list of titles reflects our commitment to bringing the best of an expanding body of literature to the environmental community throughout North America and the world.

Support for Island Press is provided by Apple Computer, Inc., The Bullitt Foundation, The Geraldine R. Dodge Foundation, The Energy Foundation, The Ford Foundation, The W. Alton Jones Foundation, The Lyndhurst Foundation, The John D. and Catherine T. MacArthur Foundation, The Andrew W. Mellon Foundation, The Joyce Mertz-Gilmore Foundation, The National Fish and Wildlife Foundation, The Pew Charitable Trusts, The Pew Global Stewardship Initiative, The Philanthropic Collaborative, Inc., and individual donors.

About World Wildlife Fund

Known worldwide by its panda logo, World Wildlife Fund is dedicated to protecting the world's wildlife and the rich biological diversity that we all need to survive. The leading privately supported international conservation organization in the world, WWF has sponsored more than 2,000 projects in 116 countries and has more than 1 million members in the United States.

THE SCIENCE
OF CONSERVATION
PLANNING

THE SCIENCE
OF CONSERVATION
PLANNING

HABITAT CONSERVATION
UNDER THE ENDANGERED
SPECIES ACT

Reed F. Noss
Michael A. O'Connell
Dennis D. Murphy

Island Press
Washington, D.C. • Covelo, California

Library of Congress Cataloging-in-Publicaion Data

Noss, Reed F.
    The science of conservation planning : habitat conservation
  under the Endangered Species Act / by Reed Noss, Michael O'Connell,
  Dennis Murphy.
        p.   cm.
    Includes bibliographical references and index.
    ISBN 1-55963-566-5 (cloth). — ISBN 1-55963-567-3 (paper)
      1. Habitat conservation—United States—Planning.   I. O'Connell,
  Michael A., 1963–   .   II. Murphy, Dennis D.   III. Title.
  QH76.N676   1997
  333.7'2'0973—dc21   97-34520
                                                                  CIP

Printed on recycled, acid-free paper ∞ ❀

Manufactured in the United States of America

10 9 8 7 6 5 4 3 2 1

lessons learned from large-scale planning exercises in Florida, California, and elsewhere.

In *The Science of Conservation Planning*, Reed Noss, Michael O'Connell, and Dennis Murphy make a strong case for habitat conservation through landscape-scale planning as the most promising means of conserving species and natural communities within the context of human activity. The authors describe the short history of conservation planning in the United States, respond to common criticisms of the approach, and provide detailed and specific biological guidelines for constructing, implementing, and evaluating conservation plans. Perhaps no group of authors is more qualified to produce a work of this kind. Noss, O'Connell, and Murphy are acknowledged leaders in the field of conservation biology and, perhaps more important, all are experienced practitioners of what Noss refers to as the art and science of conservation planning.

The immediate promise of conservation planning is that it provides a mechanism for defusing conflicts between development and species conservation. The continuing debate over the Endangered Species Act is fueled in large part by claims that the inflexibility of the Act hinders economic growth. The habitat conservation planning provision of the Endangered Species Act, properly implemented, can provide a powerful answer to those arguments. But more important, conservation planning as envisioned by Noss, O'Connell, and Murphy offers a realistic approach to preserving natural communities in settings where development is inevitable and simply setting aside protected areas is not a practical or complete solution.

Thoreau was among the first Americans to recognize the intrinsic value of wildlife and wildlands. But perhaps his most prescient insight was that, ultimately, men and women could not choose between civilization and wilderness, between human and natural communities. He believed that the ideal condition for humankind is a balance: the intellectual and cultural benefits of civilization, enriched by a complete "life in nature." On his farm in 1856, Thoreau realized that the natural community around him had not been maintained in the face of advancing development, and he deeply regretted that his own life in nature would be forever incomplete. This book offers practical conservation measures that may help us to slow and even reverse the impoverishment of the natural world around growing human communities, perhaps sparing future generations from sharing Thoreau's lament.

—William M. Eichbaum
Vice President, U.S. Program, World Wildlife Fund

FOREWORD

In March 1856, Henry David Thoreau reflected on the wildlife around his home in Concord, Massachusetts. Though he lived in a place still touched by wildness, beloved by him for its isolation and natural beauty, Thoreau perceived and mourned what was missing. "I spend a considerable portion of my time considering the habits of the wild animal, my brute neighbors," he wrote. "But when I consider the nobler animals that have been exterminated here—the cougar, panther, lynx, wolverine, bear, moose, deer, the beaver, the turkey, etc., etc.—I cannot but feel as if I lived in a tamed and . . . emasculated country. Is it not a maimed and imperfect nature that I am conversant with?"

Today, in much of the country, people are beginning to perceive, as Thoreau did, what they have lost, or stand to lose, as development proceeds unchecked by concern for the wildlife and plants—the natural communities—that share the landscape with human ones. In the burgeoning counties of the Rocky Mountains, there is a growing realization that the very fish, wildlife, and other natural amenities that prompt economic expansion in the region are threatened by that growth. Similarly, in the southwestern United States, home to the most biologically diverse deserts in the world, growth pressures are forcing conservationists, business interests, community planners, and federal, state, and local governments to recognize their impact on this most fragile of natural landscapes. In areas where high biodiversity and rapid development are already in collision, such as south Florida and southern California, intricate planning processes are under way to reconcile human and natural communities. It is conservation planning such as this that is the subject of this book.

World Wildlife Fund produced the first major report on habitat conservation planning under the Endangered Species Act in 1991. *Reconciling Conflicts under the Endangered Species Act: The Habitat Conservation Planning Experience* examined the few habitat conservation plans that were then in operation, and offered recommendations for improving the design, implementation, and monitoring of conservation plans. This new book builds on the foundation of the 1991 report, discussing in detail how the evolving principles of conservation biology can enrich the planning process, and drawing on

CONTENTS

Foreword, by William M. Eichbaum   ix

Preface   xi

Note from the Authors   xv

1. Species and Habitats   1

2. Habitat-Based Conservation Planning: A Brief History   19

3. Criticisms of Science in Habitat-based Conservation Plans   49

4. Principles for Habitat-based Conservation   73

5. Criteria for Assessing the Adequacy of Conservation Plans   111

6. A Framework and Guidelines for Habitat Conservation   155

7. Conclusions   209

Literature Cited   225

Index   239

PREFACE

In this book we review the science of habitat-based conservation planning, especially as it has been and could be applied under the U.S. Endangered Species Act of 1973. Conservation planning is really as much an art as a science. Both aspects are equally important. We concentrate on the science in this book not just because we are scientists but because we believe a strong scientific foundation is the best assurance that a conservation plan will succeed in today's crisis-driven, contentious, litigious, yet science-respecting society. It is the scientific basis of a plan that will be most intensely scrutinized by all parties in a conservation planning process, including developers, environmentalists, politicians, journalists, academics, and of course lawyers. When a plan is grounded in defensible science, it will be more immune to political interference and lawsuits and more likely to achieve its biological goals. The art of conservation planning, on the other hand, comes with experience and intuition. It has to do with synthesizing all sorts of quantitative and qualitative information—something the human brain does much better than any computer—and putting it all down on a map. The art is more difficult to put into words. And without a scientific foundation, the art has no substance.

I write this preface on behalf of my coauthors. I first became involved in conservation planning in the early 1980s when I worked for the Ohio Department of Natural Resources. When the Department showed little interest in (and, in fact, some hostility toward) my rather audacious proposals for nature reserve networks in southern Ohio, I left Ohio and pursued my work on regional reserve networks as a conservation activist, chiefly in Florida. In 1990 I began working with my first coauthor on this book, Mike O'Connell, on a World Wildlife Fund (WWF) project concerning biodiversity conservation on private lands. Mike was employed by WWF at the time, and the major focus of our project was developing habitat assessment and protection guidelines for voluntary conservation projects on corporate lands. Concurrently, Mike was a coauthor with Michael Bean and Sarah Fitzgerald on the first major report, published by WWF in 1991, on habitat conservation plans (HCPs) under Section 10 of the

Endangered Species Act. The present book, planning for which began in 1991, is a belated and elaborated sequel to the HCP report by Bean, Fitzgerald, and O'Connell.

My second coauthor, Dennis Murphy, also has extensive experience with habitat conservation planning both on private and public lands. After a long involvement with conservation plans in California, especially regarding endangered butterflies, Dennis became a member of the scientific committee, chaired by Jack Ward Thomas, for conservation of the northern spotted owl. The Thomas committee's report, published in 1990, is still considered a model for the application of scientific principles, procedures, and information to conservation planning. I began working with Dennis in 1991 as a member of the governor-appointed Scientific Review Panel for the Natural Community Conservation Plan (NCCP) for the coastal sage scrub of southern California, a panel that Dennis chaired. The NCCP process, which is promising but has generated heated controversy, tied all three coauthors together in 1995 when Mike O'Connell left his position as HCP director for The Nature Conservancy in Florida and became The Nature Conservancy's NCCP director in California.

Habitat conservation planning, especially under the Endangered Species Act, has blossomed in recent years but is extremely contentious, not only pitting environmentalists against developers and the regulated community, but also threatening to split the environmental community itself. The science applied to HCPs has varied from nonexistent to flimsy to highly sophisticated. Consistent and defensible scientific standards for plans do not exist. We have written this book in an effort to improve the scientific basis for conservation planning under the Endangered Species Act and otherwise. We hope that our work will be seen as a model for the application of conservation biology to real-life conservation problems.

Conservation biology is science in the service of conservation. It is a goal-oriented and value-laden exercise, yet to be credible it must also be as rigorous and objective as possible. We devote much of this book to exploring the role of the scientist in conservation planning; practical matters such as how to proceed with planning in the face of insufficient data and abundant uncertainty; the vexing problems of advocacy and burden of proof; and how scientists might contribute more meaningfully to policy-driven exercises such as conservation plans. Our fellow scientists, then, are a large part of the intended audience for this book. The scientists we hope to appeal to include agency biologists, academic scientists, environmental consultants, scientists employed by industry and conservation groups, and others. But we are also writing to elected officials and their staffs, to citizens who serve as environmental watchdogs of conservation plans, to developers and their contractors, to university students who wish to learn about applications of conservation science in the real world, and to everyone else who might be interested in the complex and sometimes bizarre arena of conservation planning.

Given the diversity of our intended audience, different readers will want to use this book in different ways. Those readers who are already heavily involved in HCPs, NCCPs, and other conservation plans can skim through the first couple of chapters, which are introductory and provide historical background, and begin reading seriously with chapter 3 on the criticisms of science in conservation planning. We hope that scientists and nonscientists alike will benefit from chapter 4, which reviews the principles of conservation biology that apply to conservation planning and some of the philosophical considerations involved. The most detailed examination of conservation plans—including suggestions for assessing the adequacy of plans, case studies, and a framework and guidelines for scientifically defensible plans—can be found in chapters 5 and 6. Our conclusion, chapter 7, closes with specific recommendations for scientists, environmental activists, developers and industry, and agency staff who are involved in conservation planning.

Conservation planning is not for the faint of heart. For scientists it is much easier to remain in the ivory tower and conduct basic research than to enter the messy world of biopolitics. Department chairs, search committees, promotion and tenure committees, and other scientific peers give few points to their colleagues involved in real-world conservation—in fact, it often counts against you. Courses on conservation planning and other truly "applied" aspects of conservation biology are rarely taught in universities, and students are generally discouraged from stepping into the arena of conservation planning. Universities still train students to be little professors. Those scientists who do enter conservation planning, despite the disincentives, can expect to be attacked from all angles. I have personally been called a biostitute and a disgrace to conservation biology by environmentalists for simply peer-reviewing a development plan and finding some merit in it. From the other side I have been sworn at by loggers, called an ecoterrorist by developers, and characterized as an eco-fascist by a nationally syndicated columnist. My coauthors suffer similar insults on a regular basis. But this is the real world. The human population is growing, natural areas are being destroyed, species are going extinct, and some resolution of development and conservation must be found. If conservation plans succeed in attaining their biological goals, it all will have been worth it—even if we personally are not around to see it happen.

This book languished for a few years as my coauthors and I waited to see how changes in the Endangered Species Act might affect what we are writing about and as we became busy with other projects. We owe much gratitude to Bill Eichbaum and Chris Williams of WWF for insisting that we persevere with this project, and to WWF, the Pew Scholars Program in Conservation and the Environment, and Chevron Corporation for providing funding. We thank Michael Bean, Reed Bowman, Pete Brussard, Steve Johnson, David Wilcove, and Chris Williams for their generous and helpful comments on an earlier draft. Their comments and the kind words of Barbara Dean of Island Press were so encouraging that we more than doubled the length of the text, transforming it from a long report to a reasonably sized book. Amazingly, Michael Bean and Steve Johnson were willing and able to review the entire enlarged text, and helped us correct some problematic interpretations. Any problems or errors that remain can be blamed entirely on me, as I took on the last few rounds of editing and wordsmithing. Finally, we thank Barbara Dean and Barbara Youngblood of Island Press for their expert advice and editing. This is the second book project I have worked on with Island Press, and I do not intend it to be my last.

—Reed F. Noss

A NOTE FROM THE AUTHORS

The following are definitions and uses of some key terms in this book. These terms are used variously and often loosely and interchangeably in the scientific and popular literature. Some of these terms are broadly overlapping in meaning, but we do not consider them completely interchangeable. We favor reasonably rigorous definitions, but not so strict that they conflict with standard usage.

HABITAT   The multidimensional place where an organism, population, or assemblage of populations lives; the living and nonliving surroundings. Habitat is traditionally used by biologists in a species- or population-specific sense (e.g., the habitat of the Houston toad). In this book, habitat also refers to the collective surroundings of many organisms with similar requirements (e.g., coastal sage scrub habitat). Many geographers and some ecologists speak of *habitat types* synonymously with vegetation types at one scale or another (e.g., temperate coniferous forest or sagebrush steppe). The term *physical habitat* is often used to denote a piece of environment defined by abiotic variables (e.g., serpentine barrens).

HABITAT-BASED CONSERVATION PLAN   A conservation plan that takes into consideration the overlapping habitat requirements of multiple species within the context of a functional community or ecosystem (see below). Thus, the emergent properties of the community or ecosystem (interspecific interactions, energy flow, material cycles) are considered along with the needs of the species most sensitive to human activities.

COMMUNITY   The naturally occurring assemblage of populations living in the same general place (habitat) and time. *Community* may be used to refer to all species in the assemblage or a subset, such as the plant community (e.g., spruce-fir forest) or the neotropical migrant bird community. It is generally assumed that the populations in a community affect each other to varying degrees, directly or indirectly, through trophic interactions (e.g., predation) or their influences on habitat structure (e.g., vegetation layering).

NATURAL COMMUNITY   This term, often used by The Nature Conservancy, refers to identifiable groups of organisms and their physical environments, distinguished by their biota, abiotic characteristics, or some combination of the two. For example, the natural communities of Florida include longleaf pine/turkey oak sandhills, mesic flatwoods, maritime hammock coastal strand, shell mound, freshwater tidal swamp, blackwater stream, and algal bed—along with all the associated species. Note that some ecologists would regard these as habitats!

ECOSYSTEM   A physical habitat with an associated assemblage of interacting organisms (community); a system through which energy flows and materials cycle. This definition is liberal enough to encompass plant associations, major vegetation types, and habitats defined by biological, physical, or geographic parameters and occupying a range of spatial scales (e.g., redwood forests, coral reefs, limestone caves, deep-sea thermal vents, or the Arctic Coastal Plain). To be usefill to conservationists, defined ecosystems must be discrete enough to be mapped. They must also be describable by adjectives denoting quality, such as *undammed, old-growth,* or *ungrazed.*

LANDSCAPE   A heterogeneous land area or ecosystem composed of two or more different communities, landforms, and/or land uses. Landscapes range from hundreds to millions of hectares in size and are often delineated by natural boundaries, such as watersheds or mountain ranges. Large landscapes may also be called *regions, ecoregions,* or *bioregions.* Landscapes often have characteristic patterns determined by their inherent abiotic and biotic elements and by human land uses. Landscape components are often described as *matrix, patches,* or *corridors.*

POPULATION   The organisms belonging to a particular species and occupying a specific area. Breeding populations may be distinguished from nonbreeding populations, and *source* populations (where recruitment exceeds mortality) may be distinguished from *sink* populations (where mortality exceeds recruitment). A population is considered *viable* if it is capable of sustaining itself in an area for a long time (usually measured in at least decades).

METAPOPULATION   A group of populations (or subpopulations), spatially distinct but generally connected by at least occasional dispersal. Although some biologists restrict the term to its original usage, where there is a presumed equilibrium or balance between extinction and colonization of populations on habitat patches, we prefer the more general sense of a spatially subdivided population. Thus, there are many kinds of metapopulations, most of which are not at equilibrium and some of which include populations that are virtually isolated from all others.

THE SCIENCE
OF CONSERVATION
PLANNING

1

SPECIES AND HABITATS

The U.S. Endangered Species Act of 1973, one of the strongest and most influential environmental statutes worldwide, was inspired by the recognition that human activities are driving species to extinction. Public concern about the environment in the early 1970s was centered on pollution as a threat to human health, and pollution remains the most visible of environmental issues to Americans. The "human environment" and human well-being take precedence over other species and the total environment in virtually all legislation. Our culture is decidedly anthropocentric, and we worry little about the future beyond our lifetimes. But the Endangered Species Act (ESA) is a bit different from other environmental statutes. Those who crafted the Endangered Species Act saw value in other living things and understood that human destruction of habitat threatened the existence of many species. The purpose of the Act was stated clearly: "to provide a means whereby the ecosystems upon which endangered species and threatened species depend may be conserved."

Despite its virtues the ESA is far from perfect from a biological perspective. The implied theme of saving species through the conservation of ecosystems or habitats—which might provide for truly proactive conservation—was not well developed in the ESA. In fact, the term *ecosystem* does not appear in the Act after the preliminary statement of purposes. Ecosystem conservation was an idea far ahead of its time in 1973 and remains to be firmly encoded in law today. Not only does the United States lack a national strategy to conserve biodiversity and sustain healthy ecosystems, but legal constraints on destroying habitat on either public or private land are extremely limited. Aside from wetlands regulations, zoning ordinances, and an assortment of local statutes, there are few restrictions on what private landowners can do with their lands. As we review in this book, the ESA generally prohibits destruction of the habitat of animal species listed under the Act, and this prohibition applies on private as well as federal lands. There are ways to get around this rule, however, and the mitigation required of landowners for destroying habitat of listed species often has been meager. But the bigger problem over the long term is that referred to above: the ESA, and environmental policies generally, have not encouraged proactive actions that might preclude the need to list species as endangered or threatened. Such actions fall mainly within the realm of habitat or ecosystem conservation, the subject of this book.

The meager attention to ecosystems and habitats in conservation policy is not all that surprising. These concepts are poorly understood by the general public, and even biologists seldom agree on what they mean in detail. Among many conservationists the feeling seems to be that saving species is hard enough—don't bother us with the complexity of ecosystems! But there are signs of increased interest in the idea of habitat-based conservation among scientists, legal scholars, lawmakers, and citizens on all sides of the issues surrounding endangered species. People are beginning to realize that conflicts can be avoided, or at least reduced, by fulfilling the needs of many species at once through the broad-scale conservation of habitats, and that such actions may keep some species off the endangered species list, thus reducing the regulatory burden for private landowners. To encourage this interest in habitat-based conservation and channel it along scientifically defensible lines, we have written this book.

Why Worry about Habitats?

Sustaining healthy habitats and ecosystems as a way of maintaining viable populations and preventing extinctions makes sense from a scientific standpoint (the Note from the Authors explains our use of *habitat, ecosystem,* and related terms). Simply put, if we want to save species we must protect a sufficient quality and quantity of habitats. This understanding did not arise overnight, but developed over many years of observation and research by scientists and others. The earliest humans must have observed that not only they but also other animals require food, water, and shelter to survive. Later, naturalists began to expand this body of knowledge by taking note of the particular habitat conditions under which species of plants and animals are found in Nature. Although for centuries the formal science of biology was preoccupied with naming and describing new species, some of the more perceptive individuals became intimately familiar with the habitats in which species were found and how species lived their lives. This detailed, natural historic information was central to the development, in the middle and late nineteenth century, of the science of ecology.

Ecology has always been a habitat-centered science. In 1840 the German chemist Justus von Liebig formulated the "law of the minimum," which stated that each kind of plant requires some minimal quantity of nutrients, water, or other materials to survive (Liebig 1840). Liebig's law was later extended to animals and, in 1913, it was elaborated into Victor Shelford's "law of tolerance," with the recognition that each species lives within certain bounds of temperature, humidity, soil texture and chemistry, and other factors (Shelford 1913). The concept of the ecological niche was a continuation of this line of thinking. To ecologists, the niche comprised the complete set of habitat requirements of a species—that is, the upper and lower limits of all environmental variables within which the species could survive. Many ecologists included interspecific interactions, position in the food web, and other community-level details in their characterizations of niches (Smith 1974). Although niche theory is no longer in vogue in ecology, all biologists recognize the inseparability of species and habitat, not only in terms of living requirements but also in terms of natural selection (the process of differential survival and reproduction of individuals) and the continual interchange of matter and energy between organisms and their surroundings. A distinct boundary between an organism and its environment is illusory—in a fundamental sense they are one. The centrality of the species–habitat connection in biology helps explain the confusion and outrage expressed by scientists and others over a 1994 Circuit Court of Appeals decision (later overturned by the Supreme Court) that destruction of the habitat of an ESA-listed species does not constitute "taking" of the species (Noss and Murphy 1995).

To many laypersons the habitat requirements of a species begin and end with the kind of environment you find the species in. Hence, if a spotted owl *(Strix occidentalis)* is seen in a young, second-growth forest, for example, it is concluded that spotted owls do not need old growth to survive. Being observed in a habitat and maintaining a viable population in a habitat, however, are two very different things.

The spectrum of habitats over which a species is observed varies widely in quality. Not uncommonly, an individual organism—especially of a mobile species such as most birds—wanders into an area that is clearly unfit for survival; its days in that area are numbered. A loon landing in a wet parking lot that it mistakes for a lake would be an example of this situation. In many other cases more subtle differences in habitat quality and corresponding responses of populations require intensive, scientific study to discern. In some areas—called *sources*—habitat quality is high, the rate of reproduction exceeds the rate of mortality, and the population grows or exports individuals to other areas. In some of these other sites—called *sinks*—mortality exceeds reproduction, and a population can persist only if immigrants frequently disperse in from nearby source populations. Very slight (to human observers) distinctions in habitat structure, prey populations, predators, pathogens, competitors, disturbances, and other factors may distinguish a source from a sink. The spatial configuration of sources and sinks across a landscape may ultimately determine the survival of the entire population or metapopulation.

Species-specific area requirements also determine habitat suitability. A site that appears ideal for a species in terms of habitat structure may fail to support a population, or even an individual, if it is too small. Many songbird species are "area-sensitive" and usually breed only in tracts of forest or grassland many times larger than the size of their territories (Whitcomb et al. 1981; Robbins et al. 1989; Herkert 1994; Vickery et al. 1994). The probability of finding breeding pairs or populations of area-sensitive species generally increases with the size of the habitat patch. The animals with the largest area requirements are generally large, mammalian carnivores—for this and other reasons, they are absent from many regions of North America today. For example, in the Rocky Mountains, individual, annual home ranges are on the order of 150 km^2 for black bears, more than 400 km^2 for mountain lions and wolverines, and nearly 900 km^2 for grizzly bears. The gray wolf, a social predator, uses from 250 to >2,000 km^2 per pack territory. For long-term persistence of populations of these species, wild areas on the order of 100,000 km^2 or larger seem to be required (Noss et al. 1996 and references therein). Because no single reserve in North America is this large, conservation proposals for these species must employ networks of reserves connected by regional-scale corridors. But even for species much less demanding than large carnivores, habitat needs often must be considered in terms of the constellation of patches of potentially suitable, potentially connected habitat across a large landscape, rather than site by site. Single sites or a collection of disconnected sites will often be insufficient for long-term survival.

Because of factors such as those just summarized, habitat-based conservation is more complicated than it might initially seem. It gets even more complex when we begin considering the needs of many species at once, along with the ecological processes that keep their habitats in suitable condition. The science behind the conservation of communities and ecosystems is still in its infancy, though it is developing rapidly and has made substantial progress in recent years. Despite the challenges, there is no alternative to habitat-based conservation, short of saving species only in captivity—and this is really no alternative at all.

The Problem of Habitat Alteration

Habitat-based conservation is logical and consistent with accepted scientific principles. But there is a more urgent reason to focus on habitats and ecosystems in conservation planning: natural habitats are disappearing rapidly. Scientists agree that habitat alteration is the greatest threat to species and ecosystems in the United States and worldwide (Ehrlich and Ehrlich 1981; Diamond 1984; Wilcox and Murphy 1985; Wilson 1985; Ehrlich and Wilson 1991; Soulé 1991; Noss and Cooperrider 1994; Noss et al. 1995). Habitat alteration comprises not only the direct, physical conversion of a natural area to an unnatural habitat (*habitat destruction*)—for example, rainforest converted to cattle pasture or a wetland replaced by a parking garage—but also includes breaking a large, contiguous patch of habitat into smaller patches (*habitat fragmentation*) and changes in the composition, structure, or function of an ecosystem (*habitat degradation*) (Noss et al. 1995; Noss and Peters 1995). These processes can take place on several spatial scales, from the "internal" degradation or fragmentation of a small patch of relatively uniform habitat, to the fragmentation and homogenization of entire landscapes and regions (Noss and Csuti 1994).

Habitat alteration may be caused by agriculture, logging, mining, urban development, fire suppression or other changes in the natural disturbance regime, alteration of stream flows because of dams or diversions, disturbance by off-road vehicles, heavy grazing by live-stock, pollution, introduction and invasion of nonnative species—which also have biological effects, such as predation and competi-tion—or other factors. Note that most of this destruction is not malicious; it is an unintended consequence of normal, basically legiti-mate economic activity (habitat alteration by golf course and ski resort development, off-road vehicle use, and other recreational activities is an exception to this rule). The connection between human land use and habitat alteration is difficult to break.

Fundamentally, human activities need not result in severe and irre-versible losses of biodiversity. We evolved on this Earth and, at least in some places and times, had an apparently sustainable relationship with the rest of Nature. Perhaps this relationship can be renewed. This realization is a basic premise—or hope—of conservation planning and ecosystem management. We know that, in many cases, relatively minor changes in how we extract resources, design developments and highway networks, or manage habitats, can make tremendous differ-ences in terms of impacts on species and communities. These changes must be made. But with a growing population and economy, the pos-sibility of a sustainable relationship between humans and Nature be-comes increasingly remote. With so many of us, consuming so much, our collective impact is staggering. Barring major changes in the scale and intensity of human activity, the inevitable consequence of human land use is habitat destruction, fragmentation, and degradation, which in turn result in a decline in populations of species that are unable to adapt to the new habitat or its new occupants. Population declines are usually first evident on a local scale, but ultimately widespread habitat alteration results in endangerment and extinction of species. This story has been played out for at least hundreds of years (Noss and Cooperrider 1994). We expect it to continue but are committed to finding ways to lessen the damage. This is perhaps the best conserva-tionists can ever do.

Listing of species under the ESA is one, often belated, sign that habitat alteration has gone too far. As of 31 December 1996, the U.S. Fish and Wildlife Service (FWS) and National Marine Fisheries Ser-vice listed 1,051 species (437 animals [which includes four species with dual status], 614 plants) in the United States as endangered or threatened under the ESA. Habitat alteration is considered the leading factor in the endangerment of species in the United States (Flather

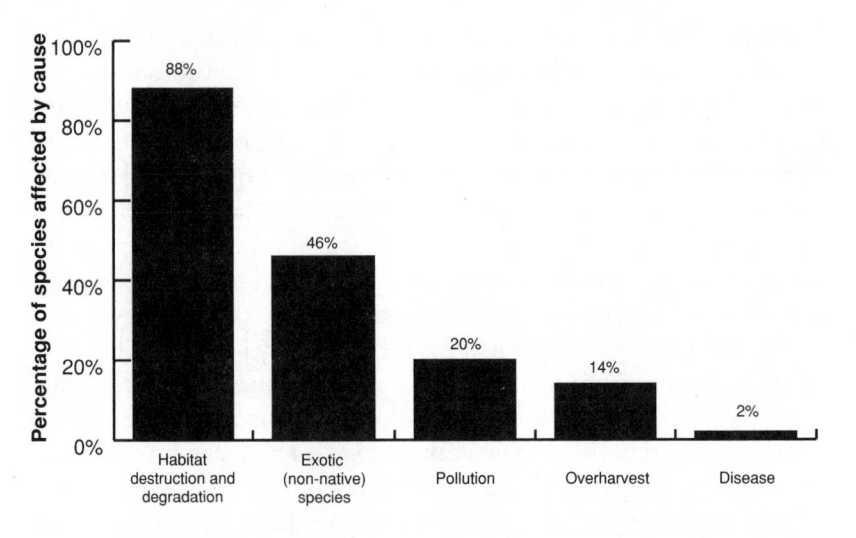

FIGURE 1.1. Causes of endangerment of species listed as threatened and endangered in the United States. A species can have more than one cause of endangerment. Note that habitat alteration is by far the greatest threat, followed by deleterious interactions with exotic species, which often invade following habitat alteration. Adapted from Wilcove et al. (1996). The data in this figure were obtained from the Federal Register and cover all species listed or proposed for listing as of 31 December 1995.

et al. 1994; Wilcove et al. 1996; fig. 1.1). More specific studies have come to similar conclusions. For example, direct habitat destruction was a contributing factor in the disappearance of three-quarters of the 27 species and 13 subspecies of freshwater fishes that have gone extinct in North America over the past century (Miller et al. 1989) and, along with introductions of nonnative species, remains the leading threat to fish species and aquatic biodiversity in general (Williams et al. 1989, Allan and Flecker 1993).

In part because habitat alteration has not abated, the list of endangered and threatened species has grown steadily since the ESA was first passed, except for a period between March 1995 and April 1996, when Congress imposed a moratorium on species listings. The list of candidate species (those being considered for listing) also increased rapidly from the inception of the ESA until July 1995, when it shrank by nearly 4,000 species with the FWS's elimination of Category 2 (C2), those candidates for which sufficient information to decide whether listing is warranted is not currently available. The decision to eliminate these species from consideration for listing was based not on scientific information but apparently on the perceived political liability of having such a large list of candidates. At the time of this writing the candidate species list contains only those species for which sufficient information to list is in the hands of the FWS, and includes 182 Category 1 (C1) species and an additional 238 species proposed for listing (i.e., well into the listing process). The burden of proof for listing is on citizens who petition the government to add species to the list, and few government funds are available for status surveys.

Because getting a species listed is influenced by politics (a classic example is the northern spotted owl, *Strix occidentalis caurina*, whose listing was delayed for several years due to factors that a federal judge determined had nothing to do with biology), a better indication of the status of species can be found in lists produced by independent scientific and conservation organizations. In particular, The Nature Conservancy ranks species in terms of their rarity and vulnerability at global and state scales. In the United States, 1,339 species (160 vertebrates, 166 invertebrates, and 1,013 vascular plants) are considered critically imperiled globally (G1); 1,831 species (157 vertebrates, 157 invertebrates, and 1,517 vascular plants) are considered imperiled globally (G2); and an additional 3,076 species (256 vertebrates, 264 invertebrates, and 2,556 vascular plants) are considered vulnerable globally (unpublished Natural Heritage Central Database, January 1996). These figures—together with others such as the number of species that have already gone extinct in the United States since European settlement—give a more accurate picture of the status of biodiversity than the official list of endangered and threatened species.

Meanwhile, the habitats of many species, listed and unlisted, continue to be diminished and degraded. Left unchecked, habitat alteration will inevitably result in the extinction of many of these species. Currently, far more listed species are declining than increasing in population size (U.S. Fish and Wildlife Service 1994; Wilcove et al. 1996; fig. 1.2). Moreover, listed species on private land are in worse peril than those on federal land. For listed species occurring only on private land, the ratio of declining to improving species is 9:1, compared to 1.5:1 for listed species found entirely on federal land. Apparently because of the reluctance of landowners to allow biologists on their properties, the FWS is unaware of the status of more than half of the species found only on private land (Wilcove et al. 1996).

But the problem of habitat alteration is still bigger. Entire eco-

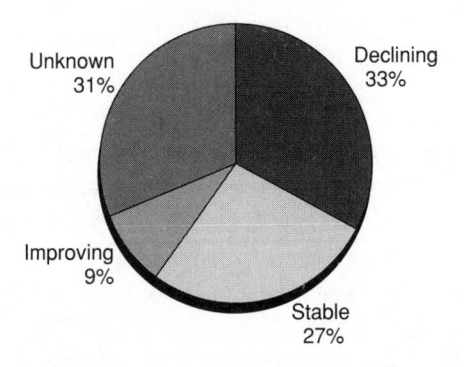

FIGURE 1.2. Status of listed species in the United States. Note that many more species are declining than improving in status, and that the U.S. Fish and Wildlife Service does not know the status of nearly one-third of listed species. The situation is more dire on private lands (see text). Adapted from Wilcove et al. (1996) and based on data from U.S. Fish and Wildlife Service (1994).

systems—habitats, communities, and landscapes—in the United States have declined greatly and, in some cases, vanished because of human activities. Research commissioned by the National Biological Service (Noss et al. 1995) and a follow-up study by Defenders of Wildlife (Noss and Peters 1995) determined that a large number of ecosystems, many of which are unique to North America, are endangered. Among the major ecosystems that have suffered substantial losses and remain at risk in the United States are those listed in box 1.1. Direct habitat destruction, fire suppression and other disruptions of natural disturbance regimes, and such secondary effects as invasion of exotic plants and animals were identified as the major threats to these ecosystems. As these ecosystems decline in extent and quality, so do populations of the species that compose them. It stands to reason that conservation strategies, including restoration, focused on entire ecosystems or landscapes will be more efficient than separate plans prepared for hundreds of individual species at potentially thousands of sites.

Habitat-based Conservation

If alteration of natural habitats is the primary factor threatening species and ecosystems, then conservation of natural habitats offers

Box 1.1.

The most highly endangered major ecosystems of the United States, as determined by a coarse analysis of extent of areal decline since European settlement, current rarity (areal extent), number of endangered and threatened species associated with each type, and level and urgency of continuing threats. Adapted from Noss and Peters (1995).

South Florida landscape

Southern Appalachian spruce-fir forests

Longleaf pine forests and savannas

Eastern grasslands, savannas, and barrens

Northwestern grasslands and savannas

California native grasslands

Coastal communities (terrestrial and marine) in the lower 48 states and Hawai'i

Southwestern riparian communities

Southern California coastal sage scrub (and associated communities)

Hawaiian dry forest (and associated communities)

Large streams and rivers in the lower 48 states and Hawai'i

Cave and karst systems

Tallgrass prairie

California riparian communities and wetlands

Florida scrub

Shrublands and grasslands of the Intermountain West

Ancient eastern deciduous forest

Ancient forests of the Pacific Northwest (including redwoods)

Ancient red and white pine forests of the Great Lakes states

Ancient ponderosa pine forests

Midwestern wetlands

Southern forested wetlands

the best hope for slowing, halting, or reversing these losses. Furthermore, those habitats or communities that have suffered greatest losses or are most vulnerable to further losses in the near future should be of highest priority for protection. This simple fact has been recognized by conservationists for centuries (Grove 1992; see chapter 2). Since the late nineteenth century the primary conservation strategy recommended by scientists worldwide to compensate for habitat alteration has been to represent examples of all natural communities in protected areas. This strategy requires, first of all, an assessment of how well communities are currently represented in reserves. In the United States the Gap Analysis of the National Biological Service (NBS, now incorporated within the U.S. Geological Service) is an example of a representation assessment. Existing data show that, in all countries examined, representation of natural communities is incomplete, and certain kinds of habitats—generally those most accessible and useful for economic activity—tend to be poorly represented in reserves. These habitats include coastal areas, riparian and bottomland habitats, fertile grasslands, low-elevation sites in mountainous landscapes, and well-watered sites in arid regions (Usher 1986; Pressey et al. 1993; Scott et al. 1993; Noss and Cooperrider 1994).

Implementation of the ESA has focused largely on habitat protection and, in some cases, restoration. After a species is listed, the ESA requires that a recovery plan be prepared for that species. These plans usually emphasize habitat conservation of some type. Unfortunately, recovery planning under the ESA is far from a success story. Recovery plans have been of marginal scientific quality and of limited effectiveness in many cases, and only about half of listed species currently have recovery plans completed. Although some recovery plans cover several species associated with the same habitat, the vast majority are single-species plans that are not well coordinated with recovery plans for other species associated with the same ecosystem. We feel there is considerable opportunity for more multispecies and ecosystem-scale recovery plans and for better coordination of recovery planning with habitat-based conservation planning.

The ESA, in Section 9, protects individual organisms of listed species and subspecies from "taking," which includes harassing, harming, pursuing, hunting, shooting, wounding, killing, trapping, capturing, and collecting. The FWS, the primary administrator of the ESA for all nonmarine species, has interpreted the intent of Congress broadly by considering habitat alteration a form of "harm" in cases where it indirectly kills or injures individuals of listed species, such as by "impairing essential behavior patterns." In 1995 the U.S. Supreme Court (*Sweet Home* vs. *Babbitt*), in overturning the Circuit Court of Appeals, confirmed that indirect harm due to habitat destruction can be considered taking and is prohibited under the Act. What kind or degree of habitat alteration constitutes harm is not spelled out specifically for most species. Furthermore, the taking prohibition does not encompass alteration of currently unoccupied habitat that may be essential to the persistence and recovery of listed species (Wilcove et al. 1996).

Two major exceptions exist to the ESA's prohibition of harming listed species. On federal lands agencies can "incidentally take" individuals of listed species if the Secretary of Interior finds that their actions do not jeopardize the survival of the species as a whole (see Sections 7 and 9 of the ESA). Similarly, under Section 10(a) of the ESA, private citizens can receive an incidental take permit if they develop a Habitat Conservation Plan (HCP) that compensates for the taking of individuals by promoting survival of the population or species in some other way. Incidental take permits are not required for plants on private lands because the ESA does not prohibit taking of plants on these lands (i.e., under the common law, wild animals are owned by no one, whereas plants are part of the land and thus owned by the landowner). Significantly, although the language of the ESA does not explicitly encourage habitat management or restoration—or the design of reserve networks—habitat-based conservation planning, done right, can accomplish these tasks and promote recovery of listed plants and animals on private as well as public lands (Wilcove et al. 1996).

| | | |

To summarize, despite the stated goal of the ESA to conserve ecosystems, conservation planning under the ESA, so far, has seldom achieved this goal or even attempted to do so. The problem of habitat alteration has not been addressed in any meaningful way. Designations of critical habitat (the habitat considered essential for recovery of each species), preparation of recovery plans, and development of HCPs have usually focused on single species and have proceeded along separate and isolated tracks. This fragmentary approach has been generally unproductive, and its scientific credibility has been questioned. Well-accepted standards for population viability analysis, reserve design, and other aspects of conservation biology are frequently ignored both in recovery plans and HCPs (Bean et al. 1991; Tear et al. 1993, 1995). Hence, conservation planning under the ESA has been criticized by many. Although we are among the critics of conservation planning, as usually practiced to date, our premise in writing this book is that change for the better is possible. We believe that the application of more rigorous, focused, well-documented, and integrated science will reduce the level of contention about conservation planning and make the goals of habitat and species conservation more readily attainable.

About This Book

Our goal is to present a framework and guidelines for applying science to regional habitat-based conservation planning. We intend plans based on our guidelines to be biologically conservative, scientifically defensible, politically realistic, and able to provide a high probability of meeting widely accepted conservation goals. We recognize that few conservation plans currently meet these criteria and that no future plan will be scientifically or biologically perfect. Nevertheless, we are optimistic that more plans will approach an optimal condition if science is applied wisely to the planning process.

The general subject of this book is habitat-based conservation, by which we mean the selection, design, and management of protected areas and multiple-use lands—both public and private—in order to fulfill conservation goals. The specific focus of our book is ESA-associated conservation planning, including single-species and multi-species HCPs, natural community conservation plans (NCCPs; currently in use in California), and related efforts. These efforts, because they correspond to Section 10(a) of the ESA, are focused on private lands.

Conservationists are increasingly concerned about endangered species on private lands. Fully 50 percent of all listed species in the United States have no known occurrences on federal lands and 78 percent of listed U.S. species have fewer than half of their known occurrences on federal lands. Also, development pressures on private lands are increasing greatly in many regions. As noted earlier, the prospects for recovery of listed species on private lands are currently far poorer than on public lands (Wilcove et al. 1996). Nevertheless, in many cases public lands play significant roles in HCPs and NCCPs and are often assumed (rightly or wrongly) to be reserves for the species and communities concerned.

Although conservation planning under the ESA is our chief focus, we intend our discussion to have much wider applicability than HCPs and NCCPs. For one thing, habitat-based conservation planning, under the ESA and otherwise, is a rapidly evolving process. We may not have HCPs in 10 or 20 years, but we hope the principles of habitat-based conservation planning that we develop here will still be relevant then. Moreover, we believe that most of what we discuss in this book is relevant now to ecosystem management and other habitat-based efforts on public lands. The science behind conservation planning should pertain equally well to private and public lands, though the applicable laws and regulations differ.

We distinguish two approaches to conservation planning under the ESA and other federal, state, local, and private initiatives in the United States: species-based conservation (for example, most HCPs), and habitat-based (community or ecosystem) conservation. Although a conservation plan for a single species could be technically habitat-based, in that it addresses the species' habitat requirements, if the plan fails to address the needs of other species with similar requirements or other aspects of the ecosystem, such as critical processes, we do not consider it a habitat-based plan (see Note from the Authors). It is important to note that we do not classify conservation plans for multiple species as habitat-based unless these species are associated explicitly through a natural community, ecosystem, or landscape. Species-based and habitat-based approaches both have the potential to meet the letter and spirit of the ESA and other federal and state laws. Each also has the potential to advance the broader goals of conserving biodiversity and ecological integrity (see Angermeier and Karr 1994; Noss 1995). However, we feel that the traditional single-species or "species-by-species" approach to conservation has many limitations (box 1.2). It is simply inefficient—not to mention costly and often biologically futile—to approach each species on an individual basis. In most cases habitat-based conservation (including ecosystem management, stripped of its political overtones) is a better approach because it allows the overlapping requirements of many species to be addressed at once from a much broader perspective—that of their collective habitat needs and interactions in the context of a process-driven ecosystem.

We caution that it would be improper to abandon species-level conservation in the pursuit of poorly defined "ecosystem" goals. Many ecosystem management proposals for public lands suffer from this problem. Knowledge of and attention to the life histories and requisites of sensitive or keystone species are essential to the effective conservation and management of ecosystems. We suggest that by more efficiently selecting species for planning and monitoring, many natural-community or ecosystem conservation goals can be accomplished (see chapter 5, box 5.1). By the same token, some of the most sensitive species, such as those with exceptionally small populations or narrowly specialized requirements, must be addressed by population-level (or even individual-level) intervention until populations have recovered to a point where management at an ecosystem level will be sufficient to prevent extinction. Therefore, species conservation and ecosystem (habitat-based) conservation not only are complementary, they must be reconciled (see box 1.2).

Following a brief survey of past efforts in conservation planning, we present some principles of conservation biology that we suggest should underlie the development of future plans. With these principles in mind we discuss criteria that can be used to assess the biological adequacy of existing and proposed plans. We conclude by presenting a general framework and guidelines for habitat-based conservation planning, which include use of population biology and demographic data in map-based hypothesis testing; adaptive management and risk analysis; use of ecological indicators; explicit consideration of ecological and evolutionary processes; application of principles of reserve selection and design; use of modern tools such as geographic information systems and spatial simulation modeling of species viability; and linking of disparate planning efforts into a unified approach.

Habitat-based conservation planning under the ESA is by nature a politicized process. We recognize that scientific decision making is one of several approaches and issues that must be reconciled during planning. But we also strongly believe that a foundation in solid, honest, and reasonably objective science (i.e., complete objectivity is impossible, but it is a valid goal nonetheless) not only can make the political process less onerous and more credible, it also can ensure the best chance of successful conservation over the long term. Even in cases where biological and ecological information is insufficient, application of the techniques, analyses, and principles described in this book can increase the probability of a plan meeting its conservation goals.

With the current movement toward providing private landowners

Box 1.2.
Conserving Species and Ecosystems?

In conservation, as in every field, people often characterize hypotheses and options as mutually exclusive. This makes it easier for them to make their point, while conveniently discarding the alternatives. Two ostensibly competing paradigms in conservation are species conservation and ecosystem conservation. At one extreme, the argument for species conservation is that each species is unique and requires individual and separate attention. At the other extreme, the argument for ecosystem conservation is that we should focus on ecosystems as organismic entities, rather than worrying about individual species. According to this argument, species themselves don't matter; it's the functioning of the ecosystem and the interactions among species that count.

We believe both extremes are ill conceived. What we call habitat-based or ecosystem conservation (see Note from the Authors) tries to reconcile species and ecosystem conservation as they are often portrayed. In truth, we are interested in conserving biodiversity at all levels of organization—from genes to landscapes. But we do not advise considering each level of organization in isolation; they must be integrated into a comprehensive plan. The ecosystem—including the collective habitat needs of all native species and the processes through which species interact with each other and the physical environment—is an appropriate target level for conservation planning and for integrating concerns from other levels of organization (Noss 1996a). Moreover, a landscape or regional scale of analysis (hundreds to millions of hectares) is usually appropriate. On this spatial scale multiple patches and the connections between the habitats and communities of interest can be considered, along with the interactions that take place among different communities in the landscape mosaic. Because popula-

and other participants in conservation plans ironclad assurances for their efforts—reflected in such recent policies as "no surprises" (i.e., an Interior Department policy stating that landowners will not be burdened with additional regulations, costs, or other requirements after an HCP is approved)—the role of science in informing the conservation planning process is even more crucial. Why? Because the margin for error has become smaller. Keeping in mind the tremen-

tions are spatially subdivided over large areas (Fahrig and Merriam 1994), and processes critical to ecological integrity often take place at these scales, ecosystems should often be defined on the scale of landscapes or regions.

Conservation biologists today generally agree that maintaining viable ecosystems is likely to be more efficient, economical, and effective than a species-by-species approach. But in recognizing the limitations of the species-by-species approach, we do not suggest that conservation planners ignore species. To the contrary, species often define the ecosystem and determine the scale at which an ecosystem should be managed. As noted by Lambeck (1997), "Although approaches that consider pattern and processes at a landscape scale help to identify the elements that need to be present in a landscape, they are unable to define the appropriate quantity and distribution of those elements. . . . Therefore, we cannot ignore the requirements of species if we wish to define the characteristics of a landscape that will ensure their retention." We feel that most species are well served by applying the coarse filter of protecting habitats and communities. Generally, the species that need special attention are those with the largest area requirements, specialized habitat needs, functional importance in the community, or greatest sensitivity to human activities (see chapter 5, box 5.1).

The problem with species conservation as often practiced, then, is not that it considers the needs of species (a plan should do that!), but that it fails to prioritize species sensibly and that it considers each species one by one. Thus, it is the *species-by-species approach*, not the *species approach* with which we quibble. Just as problematic, in terms of being piecemeal and disjointed, are planning site-by-site and threat-by-threat. All of these approaches are inefficient. Elsewhere in this book we offer suggestions for integrating concerns from many sites and many species into a comprehensive planning strategy. Such integration will not be easy, but we believe nothing else will suffice.

dous uncertainties that accompany any planning exercise that deals with natural ecosystems, scientists must be able to forecast, within reasonable confidence limits, the outcomes of alternative conservation actions. Given the fact that science is much better at explaining than predicting (Ehrlich and Ehrlich 1996), this is a daunting responsibility.

In addition to presenting an ambitious framework and guidelines for applying science to conservation planning in regions that still have the potential to maintain their full native flora and fauna, we provide recommendations for situations where major compromises are inevitable because humans dominate the landscape. In such cases conservation plans may be far from ideal, from a pure conservation biological perspective, but must still offer a high probability of meeting the biological goals agreed on by the planning participants. These goals will necessarily be more modest than in regions with more natural habitat and fewer economic pressures. Goals will have to reflect the severe constraints of the planning situation, for example, insufficient habitat to allow long-term persistence of certain target species within the region, according to the best available models. Conservation plans must be realistic, yet conservative and prudent—this is a difficult, though not impossible, mix. The meaning of conservatism and prudence in this sense is to reduce, as much as possible, the foreseeable risks to populations and communities. We emphasize that unless a plan can provide reasonably solid assurances or mechanisms to meet stated conservation goals, it should not be characterized as a "conservation" plan.

2

HABITAT-BASED
CONSERVATION PLANNING:
A BRIEF HISTORY

If habitat destruction is what imperils species, then habitat protection—conservation of the ecosystem—provides for their salvation. Somewhere in the blurry history of human concern for other living things, the need for habitat conservation was realized. But this realization came slowly and is still far from complete. Until very recently direct killing—overhunting for subsistence, the market, or sport—was the chief human threat to wild species. Animals were the victims, and as far as we know, few plant species were over exploited to the point of extinction. For most of the four million years of hominid evolution, we were not capable of dramatically changing the face of the landscape. But in the few thousand years since extensive agriculture began and, more intensely, over the three centuries since the industrial revolution, human capacity for altering habitats has escalated quickly. Few regions on Earth are now secure from our chain saws, bulldozers, earth-movers, plows, dredges, and other equipment. Perhaps even fewer regions are secure from the exotic species we transport everywhere we go, intentionally or not. Today, plants, invertebrates, and small vertebrates, including many species yet to be described by science, are as threatened as the large mammals and birds we have sought for meat, feathers, and furs or killed because we feared them as predators or competitors.

Because habitat alteration has only recently in human history become a threat to biodiversity, we are not well prepared psychologically, sociologically, or politically to address the problem. Indeed, and surprisingly to biologists, the phenomenon of habitat alteration leading to extinctions of species is not understood by much of the public. Many urban dwellers think of hunting as still the primary threat. Their compassion is for individual animals, not populations or species. Their reasons for wanting to protect animals are purely emotional, and animals are valued in terms of their sentience or similarity in consciousness to humans. Other organisms, such as plants and invertebrates, which are not considered sentient, are of little concern. Hence the extraordinary growth of the animal rights movement in recent years and the striking lack of a corresponding habitat conservation movement based on biological goals. Although the rapid growth of land trusts in recent years is encouraging, most land-trust purchases are made to protect "open space" for aesthetic and economic reasons, not to protect habitat for species.

"One of the penalties of an ecological education," Aldo Leopold wrote, "is that one lives alone in a world of wounds" (Leopold 1953). The wounds inflicted by habitat destruction, fragmentation, and degradation are invisible to many people. Few recognize that the construction of the subdivision they live in, the roads they drive on, the logging that produced pulp for the newspaper they read, the grazing that provided the hamburger they eat, the house cats they let roam and kill the neighborhood songbirds, the ATVs they ride recklessly across the beaches and deserts, and the alien weeds that proliferate in disturbed sites and invade natural areas, affect biodiversity more than all the guns and traps in the world. Fewer still realize that very subtle changes in habitat conditions can spell life or death for populations of species sensitive to such changes. Further, our political and religious leaders have yet to address in any substantive way the ultimate source of habitat alteration—human population growth and overconsumption.

Despite these discouraging signs we believe that the basic connection of organisms with habitat is apparent to most people, and with proper education, more people will come to understand that such things as the quality, size, and configuration of habitat patches determines their suitability for species. With education people will notice the wounds and want to see them heal. As noted earlier, interest in habitat conservation is increasing in many sectors of society, including those people who have the most influence on conservation decisions. Environmentalists and developers alike understand that the environment remains an important issue to Americans and that intelligently designed networks of protected areas are necessary to avoid the crises generated by last-ditch efforts to save individual species. There is an incipient habitat conservation movement in America, and we are pleased to be part of it.

Some understanding of how habitat-based conservation planning evolved to its present state is necessary in order to make further progress. We do not intend to review, in this chapter, the entire history of protecting or managing habitats for conservation purposes. Most of our focus is on habitat conservation in the era of the ESA and on conservation planning for private lands and mosaics of private and public land. It is useful to look to the past, however, for lessons about what kinds of policies and practices have fulfilled their conservation objectives and what kinds have not. Starting with the best tools we have in hand at the moment, we can build better programs.

Early History of Habitat Conservation

Ever since humans began hunting and gathering, certain kinds of species—generally those that tasted good and were easy to catch—became vulnerable to exploitation. As the human population expanded and as weapons improved and hunting became more efficient, populations of some species declined and more than a few were apparently hunted to extinction near the close of the Pleistocene (Martin and Klein 1984). Extinctions of vertebrate species escalated over the past few thousand years as humans colonized more of the Earth's surface. For example, prehistoric human activities on islands in the tropical Pacific, especially hunting but also habitat alteration and introduction of nonnative mammals, led to the extinction of perhaps 2,000 or more species of birds (Steadman 1995). In North America, extinction rates increased markedly after European settlement, and massive alteration of natural landscapes began 250 years ago, resulting in the loss of perhaps 217 full species of plants and 71 species and subspecies of vertebrates (Nature Conservancy 1992; Russell and Morse 1992); of the presumably extinct plants, 122 were in Hawai'i and 95 inhabited the mainland. Most of the vertebrates that have gone extinct in North America since European settlement were victims of overhunting, but as noted in chapter 1, habitat loss has recently become the primary cause (Diamond 1984, Noss et al. 1995).

Just when some enlightened people first became worried about the loss of species and started to devise conservation plans is unknown. Plato expressed concern about deforestation and loss of topsoil in Greece two thousand years ago. At about the same time, forest reserves and game laws protecting certain species of vertebrates were established in India (Wright and Mattson 1996). By the mid-eighteenth century biologists and naturalists in Europe had begun to worry about habitat destruction and extinctions on tropical islands being exploited for their natural resources (Grove 1992). By 1713 trees were protected on the island of St. Helena in the South Atlantic; forest reserves were established on the West Indian islands of Tobago in 1764 and St. Vincent in 1791. The Tobago reserves covered 20 percent of the island, more than virtually any country has in protected areas today. These reserves were established primarily to protect climatic conditions and watersheds, but preventing extinctions of native species was apparently also a goal (Grove 1992).

In the temperate zone early protection efforts focused entirely on game species that were being depleted by overhunting. Closed seasons on deer were established in Rhode Island in 1639 and Massachusetts in 1694; the Massachusetts law was followed closely by the establishment of a bounty on wolves, demonstrating the prejudiced nature of early species protection measures (Noss and Cooperrider 1994). It was not until 1868 that protection laws for nongame birds were established in Britain (Grove 1992), and in 1900 the Lacey Act in the United States prohibited transport of illegally taken game and wildlife parts. Other important advances in early U.S. conservation history were the Yellowstone Park Protection Act of 1872, the National Park Organic Act of 1916, and the Migratory Bird Treaty with Canada (1916) and Mexico (1936). Notably, virtually all of the early conservation measures were regulatory and prohibitory; few were habitat-based measures, as we define them. Even the establishment of Yellowstone and other early national parks had little to do with biology; the parks were established mostly for their geological wonders and spectacular scenery (Runte 1987).

Many or even most of the achievements in conservation history over the past few centuries were initiated by scientists and amateur naturalists—elite members of the public who were interested in Nature for largely aesthetic and intellectual reasons—yet seldom were protective actions taken by governments until it was understood that their economic interests were directly threatened (Grove 1992). Similarly, the growth of the science of ecology and the increased participation of scientists in public policy in the 1960s and 1970s were closely tied to the enactment of our strongest and most science-based environmental laws, especially the National Environmental Policy Act and Endangered Species Act of 1973 and the National Forest Management Act of 1976. Decision makers evidently began to understand the ecological message that species and processes interact in ways that determine the health of an ecosystem and that human society is dependent on healthy ecosystems. Or alternately, enough of the public understood this message that the politicians could no longer ignore their concerns. In any case, at this time in our history the protection of species became linked to human well-being, including economic well-being. Despite a lack of detailed understanding of the issues by the general public or its political leaders, a general appreciation of the human–environment linkage is difficult to deny.

Habitat Conservation under the Endangered Species Act

The enactment of environmental laws in the United States and of similar measures in other countries, accompanied by a general increase in awareness of the global extinction crisis, led to the explosion of professional interest in applied ecology in the 1970s and conservation biology in the late 1980s and 1990s. Many topics of interest in conservation biology today, such as population viability analysis, reserve selection and design, and ecosystem management, have as much been driven by environmental policy as they are drivers of environmental policy. The results of this interaction have been positive, in that knowledge about species and ecosystem viability has expanded exponentially in recent years, with many practical applications. But unfortunately, conservation science has advanced more quickly than policy in recent years. Environmental policy now lags significantly behind scientific understanding, probably more so than at any time since the 1960s. With this book we try to bring one aspect of environmental policy—habitat-based conservation planning—up-to-date with current thinking in ecological and conservation science.

Although habitat-based conservation planning as we know it today is a recent phenomenon, arising in large part from implementation of the 1973 ESA, most conservation laws and policies through history have involved some level of planning to ensure that the biological elements of interest were conserved (Bean 1983; Lyster 1985). For example, under the National Forest Management Act plans for national forests are written and revised every few years in order to guide forest management in ways that ostensibly are sustainable and do not threaten the viability of species. Similarly, habitat management for conservation purposes is not especially new. Wildlife managers have been manipulating habitats for game species for a long time, and at least since Aldo Leopold's classic game management text (Leopold 1933), they generally have been doing so on the basis of sound ecological principles (though not without sometimes undesirable consequences, such as edge effects, for nongame species) (Noss and Cooperrider 1994).

Nevertheless, the ESA ushered in a new era of conservation planning that extended well beyond wildlife management and public land management traditions. The ESA set up an interesting dilemma: species listed as threatened or endangered were protected from all harm. This meant that not only direct actions, such as shooting a bald eagle *(Haliaeetus leucocephalus)* were prohibited, but also that indirect actions with no intent to cause harm, such as a timber sale or construction of a hydropower project or housing development, were similarly forbidden if they affected listed species. Thus, business as usual was no longer possible either on public or private lands. (However, as noted in chapter 1, the protection from harm and other "taking" of listed species on private lands does not extend to plant species.)

Federal agencies faced an added set of duties with respect to their actions. Through a process of "consultation" required by Section 7, the actions of any federal agency (or private actions funded by an agency) would be evaluated by the U.S. Fish and Wildlife Service or National Marine Fisheries Service for its impacts on listed species. The project might then be modified to minimize the impact to the species or stopped altogether if the activity would place the species concerned in jeopardy of extinction.

The Section 7 process has been the subject of much opinionated discussion since its inception, from both conservation and business perspectives (Thornton 1991; Yaffee 1991). Despite accusations to the contrary, statistics reveal that the consultation mechanism of Section 7 has caused little economic disruption. Of approximately 195,000 federal activities reviewed by the FWS from 1987 to 1995, 5,594 were determined to affect listed species adversely, thus triggering formal consultation. Of those 5,594 consultations, only 607 resulted in jeopardy opinions and only 100 projects were stopped; all but 13 of these were timber sales in the Pacific Northwest (Barry 1991; Jost 1996; C. Williams, pers. comm.). Many conservationists argue that more jeopardy opinions should have been issued and more projects stopped;

but on the other hand, the minor impact of the ESA on the national economy provides a strong argument for not weakening the Act. Thus, proponents of Section 7 cite these figures as evidence of the effectiveness of the program, while its critics believe that these statistics reveal a general trend of permissiveness regarding impacts to listed species or, conversely, that they don't reveal the true costs of the Section 7 process (i.e., the revenues lost while projects are delayed during consultation are not accounted for). Regardless of how this debate is resolved, Section 7 has provided a way for those involved in federal activities to address incidental impacts on species.

Until the 1982 revisions to the ESA, nonfederal activities—in particular, private development that did not require federal permits—had no similar means for resolving conflicts with the needs of listed species. Although the ESA has never been aggressively enforced, land-clearing activities in connection with construction or agriculture could, at least in theory, result in fines or jail terms if death or injury to a listed species occurred as a result of such activities (M. Bean, pers. comm.). It was with the intent to resolve this theoretical dilemma, which generated a wave of private-sector opposition to the ESA, that habitat-based conservation planning under the ESA on private lands began in earnest. Such planning is dependent on landowners having adequate incentives to participate (see box 2.1) and, especially given the current political climate, must respect private property rights (see box 2.2).

Building on a pilot project on San Bruno Mountain in the San Francisco Bay area of California, Congress in 1982 amended Section 10 of the ESA (see Bean et al. 1991; Beatley 1994). Section 10(a) enabled a process analogous to federal Section 7 consultation whereby nonfederal parties, including local governments and private landowners, could move forward with development and other projects that resulted in the incidental take of listed species if they prepared conservation plans, which came to be known as Habitat Conservation Plans, or HCPs, that mitigated the effects of the incidental take. Among other requirements HCPs were supposed to have clear goals and secure funding. It has been suggested that the intent of Congress in Section 10(a) was to pursue a "no net loss of survival probability" strategy for resolving conservation and development issues, where individual HCPs left a species no worse off than it was before the activity (Bean et al. 1991).

Unfortunately, in recent years it has become apparent that the legal

Box 2.1. Incentives to Conservation on Private Lands

Private lands are an essential component of any strategy for biodiversity conservation. Approximately 60 percent of the land in the United States is privately owned. More important, those private lands, in most regions, comprise the most productive and biologically diverse habitats. For example, the vast majority of sites with permanent water in arid regions of the West are privately owned. According to the U.S. General Accounting Office (1994), more than half of the species protected by the Endangered Species Act have more than 81 percent of their habitat on nonfederal land. Many states, such as Florida, Texas, and Hawai'i, that have a disproportionate share of listed species, also have a high proportion of nonfederal land (Wilcove et al. 1996). Further, species that are found exclusively on private land are faring much worse than their counterparts on federal land (Wilcove et al. 1996).

At least part of the difficulty in achieving conservation on private lands has been attributed to the prescriptive, regulatory approach to conservation exemplified by Section 9 of the Endangered Species Act (Keystone Policy Center 1995). Many landowners justifiably claim that they are being asked to bear an unfair share of the burden of conservation, since by the simple fact that their land still harbors rare species they are among the least culpable for the predicament of those species. Some observers have even begun to note the effect of perverse incentives to degrade or destroy habitats as a way to avoid legal obligations created by the Act (e.g., Baden and O'Brien 1993). As the debate over endangered species policy has raged, more and more attention is turning toward incentives to conservation on private lands as a means to get ahead of the prohibitions that are the strands of the Endangered Species Act safety net. An in-depth discussion of private incentives to conservation is beyond the scope of this book. However, it is useful to highlight some of the categories of incentives that are under exploration.

Most proposed incentives fall into two general categories: financial and regulatory. Whereas several regulatory incentives have been used successfully, virtually none of the financial proposals and policies has been widely employed or enacted in legislation to date. Financial incentives encompass a wide range of options, including tax relief for easements or donations, tax credits for positive management practices, estate tax reforms, land exchanges, bonuses or bounties for attracting and managing for rare species and habitats, trust funds, and outright compensation. Tax incentives have been a popular topic for discussion in

BOX 2.1. INCENTIVES TO CONSERVATION ON PRIVATE LANDS   27

many forums and have resulted in several creative proposals (Fischer and Hudson 1993; Keystone Policy Center 1995). Federal income and estate taxes are major obstacles to private landowners who wish to continue or promote stewardship of private lands for conservation. In particular, removing the $600,000 ceiling on deductible inheritance for those who maintain conservation land has been cited as a way to avoid the development or liquidation of natural resources and property. This could have a dramatic effect in areas such as the Red Hills region of southwest Georgia and north Florida, where many heirs have been forced to harvest some of the few remaining virgin longleaf pine stands that have been in families for generations, in order to pay estate taxes. In many regions high estate taxes force heirs to subdivide their properties—ranches become ranchettes—thereby reducing their conservation value.

Income tax relief has been suggested as another means to provide both individual private and corporate landowners with incentives to protect habitats. Tax credits for permanent conservation easements or donations of title, and even tax reductions for such management practices as agricultural activities that promote or encourage conservation outcomes could have wide-ranging benefits to many species. For several years the California legislature has been considering a tax proposal that, in partnership with the federal government, would potentially provide several billion dollars in credits to private landowners in exchange for donations of land or easements.

Other, less broad financial proposals include such things as paying owners rewards for finding endangered species on their land (O'Toole 1990). This incentive would simply provide a cash bonus for private landowners who attract rare species or maintain critical ecosystems on their property. Stroup (1992) has suggested, for example, that the federal government pay landowners $10,000 for every breeding pair of northern spotted owls *(Strix occidentalis caurina)* that occurs on their property. If payment were made for each of the roughly 350 pairs of owls that the U.S. Fish and Wildlife Service estimates are found on private property in Washington, Oregon, and California, the total cost would be $3.5 million, nearly two-thirds less than the $9.7 million the federal government spent on protecting the owl in 1990 (Anderson and Olsen 1993). Of course, $10,000 is little incentive for owners of old-growth coast redwoods, for example, where each harvested tree may be worth tens of thousands of dollars. Yet the concept of financial rewards for maintaining species may be worth exploring as an incentive for some situations.

Some have proposed the controversial idea of outright compensation to private landowners whose activities and land uses are restricted by the presence of legally protected species. While a straightforward proposal, this suggestion has not been able to pass the political test of balance and equity between the public good and private benefit. Simply the magnitude of funds required for such a venture makes it economically unfeasible. In addition, the effect of this approach would most likely lead to less application and enforcement of regulation in the first place, as opposed to providing a true incentive for private property owners to take positive action on behalf of conservation and avoid the consequences of regulation. Nevertheless, the Congress and most state legislatures entertain proposals for regulatory takings compensation nearly every session and some have passed.

The second broad category of incentives for conservation on private lands is regulatory. Proposals are numerous, but nearly all involve increasing long-term regulatory certainty and streamlining the process of compliance. As we have noted, assurances regarding future costs and obligations are perhaps the most important factor in bringing private landowners to the table and keeping them there in most conservation planning exercises (Dwyer et al. 1995b). Being able to predict future requirements is an extremely powerful incentive that for most landowners and resource users translates easily into direct financial benefit.

A number of the incentives in the regulatory category can be achieved through administrative reform, and the FWS has been pursuing several in recent years (many can be found in the Federal Register since 1993). Perhaps the most definitive, and not coincidentally the most contentious, has been the "no surprises" policy promoted in 1994 by the Clinton administration through the Department of Interior (see discussions elsewhere in this book). This policy did not undergo Federal Register review, one of several reasons for its controversy in the environmental community. In short, the no surprises policy says that under an approved conservation plan, if future modifications are necessary that require additional contributions of land or money, the public at large will be responsible for these costs. The simplicity and permanence of this policy has led to some consternation, particularly over the lack of a clear source for the public share. Controversy notwithstanding, in regions such as southern California, where nearly every parcel of undeveloped land is home to one or more listed species and is also involved in one or more conservation planning processes, this policy has made the difference between active participation in conservation by private landowners and directing their resources instead toward campaigns to eliminate the law altogether.

Streamlining the process for obtaining incidental take permits is also an attractive regulatory incentive for private landowners, although we caution that any streamlining that undermines the scientific foundation of conservation plans is very dangerous, as it may have deleterious, population-level consequences.

Box 2.1. Incentives to Conservation on Private Lands   29

The Section 10(a) habitat conservation planning process has proven lengthy and expensive for many participants, often due to inefficiencies in review and action on the part of agency staff. To address this issue the Service recently published a handbook of guidance for habitat conservation plans that is intended to define more clearly the administration of the HCP program and eliminate unnecessary problems and delays. For example, the guidance separates conservation plans into several categories that receive differing levels of scrutiny and have varying processing timelines depending on their impact on the focal species. Equally as important as a regulatory incentive, if not more so, is the consolidation of environmental permitting requirements, which in turn provides long-term assurance for a variety of conservation activities. The value of fulfilling regulatory obligations for protected species, wetlands, and open space and parklands in a single package is a strong positive attractant for a private landowner.

The incentive of getting ahead of the regulatory process for private landowners frequently means more than just complying with the requirements of currently protected species. Long-term predictability in many cases depends on avoiding the consequences of future listings. Consequently, in recent years there has been a move toward planning and receiving assurances for unlisted or candidate species. These prelisting agreements have become a popular proposal for legislative reform (Keystone Policy Center 1995), but they are hindered from the scientific perspective that many unlisted species are unprotected because little information is available to determine their status. Precluding listing and allowing taking of rare species that we know little about has obvious dangers. Many of these species, however, may be addressed reasonably well through a habitat-based approach and the application of some of the techniques described elsewhere in this book. We agree with the frequently recommended approach of planning for these species as if they were already listed, but this implies that conservation and monitoring requirements for these species be as rigorous as for listed species.

Also in the category of regulatory avoidance incentives is a recently promoted tool called the "no-take" agreement. In this strategy landowners negotiate an agreement with the FWS on a conservation plan or series of activities that are determined to result in no incidental taking of protected species over time. In return, they are given the assurance that if the specified activities are maintained and do not cause detriment to the focal species, the landowner will remain free from threat of enforcement under the Endangered Species Act. This new tool has not been implemented widely enough at this writing to enable an objective assessment of its conservation value or problems. One example of such an agreement is the massive plan for the red-cockaded woodpecker

(Picoides borealis) negotiated between the Georgia Pacific Corporation and the Department of Interior in 1993 on the company's entire land holdings in the southeastern United States.

Among the most creative new proposals emphasizing incentives for private land is the "safe harbor" program created by the Environmental Defense Fund and adopted as an ESA administrative reform by the FWS in 1995 (Wilcove et al. 1996). This policy is intended to reward good land stewardship and nullify the perverse incentive to destroy habitat. It encourages property owners to improve their land for rare species beyond what the law requires and, in exchange, benefit from being able to undo those improvements down to a predetermined "baseline condition" without being penalized. In such landscapes as the sandhills longleaf pine ecosystem of North Carolina, where landowners were cutting pines before they became suitable for red-cockaded woodpeckers, to avoid being caught with woodpeckers on their land, this policy has been a powerful incentive to conservation in this region—for example, some 20,000 acres of sandhills in North Carolina are enrolled in safe harbors. The policy is particularly applicable in situations like forestry and agriculture, where long-rotations or periodic or low-intensity resource use are the standard. It is unlikely to work well in regions such as urbanizing areas, where permanent conversion is the typical impact.

Concerns have been expressed by some biologists about habitats enhanced under safe harbor agreements drawing woodpeckers or other listed animals off nearby occupied sites, only to be extinguished later—thus possibly becoming population sinks. At this time such concerns are purely theoretical. Source-sink dynamics are complex and poorly understood. A site may change back and forth from source to sink depending on management, periodic natural disturbances, or other factors. Today, without active management, there are no long-term source habitats for red-cockaded woodpeckers and many other disturbance-dependent species. If safe harbor agreements increase the amount of land being managed favorably at any given time, they are likely to decrease some threats to the species, at least in the short term. The potential benefits of safe harbor agreements include the creation of habitat suitable for dispersing juveniles to occupy; slowing (or reversing) habitat fragmentation; sustaining currently occupied habitat (sandhills safe harbor participants commit to manage their baseline habitat as well as new habitat); avoiding or postponing demographic and genetic bottlenecks (if safe harbor agreements produce a net population increase, even temporarily); and reducing the risk of a catastrophic storm destroying a population (assuming that as a result of safe harbor agreements the population is larger at the time of the storm than it would be without safe

BOX 2.1. INCENTIVES TO CONSERVATION ON PRIVATE LANDS   31

Red-cockaded woodpecker habitat: well-spaced, mature longleaf pines. Incentive pro-grams are being explored that may encourage private landowners to maintain habitat in this condition, rather than cutting pines at a younger age or, worse yet, converting lon-gleaf pine stands to plantations of slash or loblolly pines—as seen here in the back-ground—which are unsuitable for red-cockaded woodpeckers and many other charac-teristic but declining species of the longleaf pine ecosystem. Photo by Reed Noss.

harbor). It is premature to predict whether the potential short-term benefits will be sustained over time. Populations of listed species in landscapes containing safe harbor sites should be monitored carefully. Notably, safe harbors will have its greatest conservation benefit when landowners who improve habitat on their lands elect not to downgrade it later (M. Bean, pers. comm.). At the least, safe harbors is a way of "buying time" until more comprehensive conservation strategies for listed species can be implemented. The applicability of the safe harbors concept to species and habitats not requiring periodic disturbance or ac-tive management is untested. Further, concerns have been expressed about po-tential abuses of the safe harbor policy resulting from vague standards for base-line habitat and conservation benefits in mutual agreements between the secretary of interior and landowners.

The need to address conservation issues earlier in the game and to avoid the adverse consequences of regulation has spawned great creativity in developing incentives. Unfortunately, the stalled debate over the ESA has prevented many of them from being adopted, other than through administrative action. Combined with a clear, firm backstop of prohibitions designed to prevent extinction, the incentives highlighted here and others yet to be devised may provide a means to achieve far greater protection of imperiled species than would regulation alone. We urge that all potential incentives be scrutinized carefully in terms of their potential to contribute to the fundamental goals of the ESA.

basis for the "no net loss" interpretation and for interpretations that HCPs should actually contribute to recovery is weak (M. Bean, pers. comm.). The FWS's own draft handbook on HCPs is unclear on the standards for HCP approval, stating that "ideally, applicants should be encouraged to develop HCP programs that are consistent with or contribute to recovery plan objectives for affected species; [but] at a minimum, an HCP should not preclude recovery" (USFWS 1995). Hence, there is a range of possible standards for HCPs, from strong plans that actively contribute to recovery, to no net loss, and to net loss that does not preclude recovery in the future. Even this may be a somewhat generous reading of the language of the statute, which says that the impacts of incidental taking must be minimized and mitigated only "to the maximum extent practicable." By focusing on practicality, which implies an interest in what a landowner can afford rather than what a species can bear, Section 10(a) is rather anemic from a conservation perspective. Most HCPs are also scientifically deficient. For example, mitigation requirements are often arbitrary and lack an empirical foundation in the life-history requirements of species (Bingham and Noon 1997).

It is critical to note at this juncture that we make a strong distinction between small-project, single-landowner, low-impact permits commonly issued under Section 10(a) and regional, cooperative, land-use planning processes that are truly proactive and have greater potential to contribute to recovery of listed species and prevent the need to list additional species. Congress seemed to recognize the possibility of both types of plans and provided a mechanism for handling them both. We suggest, however, that the 1982 Conference Committee Report discussing HCPs reveals a preference for large-scale planning exercises both for their biological benefits and for their economies of scale. In terms of conserving species and ecosystems, large-scale plans have much more to offer. Unfortunately, as is often the case with ESA

BOX 2.2. PROPERTY RIGHTS AND ENDANGERED SPECIES ACT   33

Box 2.2. On Property Rights and the Endangered Species Act

Accepting the premise that the only effective means of saving species is to save their habitats has an essential implication. The Endangered Species Act, which seems to be a straightforward law designed to protect wildlife, in practice becomes a law that regulates land use. With property rights serving as a cornerstone of the new conservative political contract, the ESA has become a, if not the, primary target of those who would weaken this nation's environmental laws (Dwyer et al. 1995a and 1995b).

The recent Supreme Court decision in *Babbitt* v. *Sweet Home Chapter of Communities for a Greater Oregon* (1995) upheld the FWS's broad definition of "harm" to endangered and threatened species as appropriately including actions where habitat modification "actually kills or injures wildlife by significantly impairing essential behavior patterns." This biologically responsible decision adds fuel to a supercharged property rights debate that promises legal challenges that could potentially undermine implementation of the Act on private land.

Central to any discussion of property rights and the ESA is the concept of "eminent domain," which gives government the power to condemn private property if there is an overriding public need for the land. The Fifth Amendment of the Constitution checks that condemnation power by requiring the government to fairly compensate a landowner for property so taken. Legal scholars are now asking whether regulation of private land to prevent "harm" to a protected species constitutes a regulatory "taking" of property under the Fifth Amendment.

Five Supreme Court cases now serve as the definitive modern planks in the deck of regulatory takings law. While these cases concern neither the Act nor wildlife, the rulings illustrate an emerging standard of law that some scholars believe may affect future takings cases, if such are brought to challenge the ESA.

In *Keystone Bituminous Coal Association* v. *De Benedictis* (1987), the court found that land-use regulation at issue in this case did not unjustly deprive the property owner of economic use of his land because the regulation substantially advanced a legitimate state interest. This case may be applicable to the ESA because the protections afforded to species under the Act can diminish the value of land.

In *First Evangelical Lutheran Church of Glendale* v. *City of Los Angeles* (1987), the Court awarded damages in compensation for land-use regulation it believed went too far. A potential outcome of this decision is that government will be more hesitant to enact regulations in situations where lawsuits demanding compensation might be filed, such as with endangered species protection. Furthermore, the plaintiff in this case prevailed despite a failure to exhaust all administrative remedies before seeking redress from the courts, as previously had been required. Exhaustion is viewed as one of the reasons that property takings claims have not been made against ESA actions, since implementation of the Act involves a long regulatory process with many points for administrative review.

In *Nollan* v. *California Coastal Commission* (1987), the Court required that the mitigations demanded of landowners have essential connections to the impacts of development. The need to establish such a nexus may confound ESA implementation in the future, because the scientific data necessary to establish a tight relationship between land development and remedies for impacts are difficult to establish.

In *Lucas* v. *South Carolina Coastal Commission* (1992), the Court found that if a regulation denies all economic or productive use of property, then a taking has occurred. Since HCPs allow landowners to make partial productive use of land even if some "harm" is done to some individuals of a listed species, the ESA seems not particularly vulnerable to this finding. But the Court did not clarify what level of loss of investment-backed expectation would constitute a compensable property taking. In addition, the Court ruled that a landowner need not be compensated for a taking of land if regulation falls within the scope of state nuisance law. However, ESA wildlife protections may not qualify under most tra-

implementation, practical application of the law has led to a proliferation of small-project permits. Section 10(a) has become more of a permitting process than a mechanism for ambitious, proactive, regional conservation planning.

The FWS has invested significant resources in the HCP program. As of August 1996, 190 incidental take permits had been issued (each with its own HCP) and about 200 HCPs were in development throughout the country, most for single species and a few for multiple species or natural communities (FWS, unpublished data). Most of these HCPs, especially the earlier ones, were for planning areas of less than 1,000 acres and were single-landowner plans that even collec-

Box 2.2. Property Rights and Endangered Species Act   35

ditional nuisance law, which is designed to prevent threats to public health, safety, and welfare.

In *Dolan v. City of Tigard* (1994), the Court resolved a question left open in Nollan concerning the degree of connection required to establish a nexus between development and mitigation. The Court now requires government to make an individualized determination that a required mitigation is related in both nature and extent to the impact of the proposed development. Unfortunately, implementation of the ESA is often characterized by a good deal of uncertainty about the nature of the biological protection required to assure species survival, even though the extent to which development affects wildlife may be relatively certain.

Because with each election cycle government seems to be growing less willing to fund programs that would compensate landowners for economic value lost to species protection, it seems inevitable that the tension between imperiled species and property rights will find its way to the highest court. However, creative engagement by all involved might put off that day well into the future. Capitalism unfettered is quite blind to species in need, but increasingly, private landowners and developers find conservation a good business strategy that offers real and tangible benefits, from positive public relations to enhanced aesthetic values that can increase land values. Government itself can go far to avoid a Supreme Court showdown by providing incentives for constructive negotiation—offering reasonable and scientifically valid assurances that a deal is a deal, planning on spatial and temporal scales that satisfy the needs both of species and the regulated community, and rewarding good deeds, not just threatening to punish those who would injure species.

tively contribute little to the survival and recovery of the species concerned and less to the overall federal endangered species program (USFWS 1995). For example, one HCP permit was issued in Florida for 0.5 acres of habitat for the Florida scrub jay *(Aphelocoma coerulescens)*. The FWS, particularly under the Clinton administration, has begun a shift toward HCPs involving multiple species and larger areas. Of the 200 HCPs under development as of September 1996, 132 cover areas of less than 10,000 acres, 25 exceed 10,000 acres, another 25 exceed 100,000 acres, and 18 exceed 500,000 acres (FWS, unpublished data). We recognize the occasional value in using the Section 10 mechanism to resolve small-area problems. However, these permits should be treated quite separately from regional habitat-based conservation plans because they generally have little potential for providing broad and lasting benefits for biological conservation. Most of the references to habitat-based conservation plans in this book connote regional planning efforts.

A more serious problem is the inadequate mitigation requirements for HCPs, which collectively have the potential to undermine recovery objectives. As an example of this problem Wilcove et al. (1996) report the disturbing case of the Red Oak HCP in Louisiana, in which the Red Oak Timber Company owned 1,016 acres of forest land that contained two groups of red-cockaded woodpeckers occupying 137 acres. After logging all of the forest except the small area occupied by the woodpeckers, the company sought a Section 10 permit from the FWS to log the rest. Their request was approved by the FWS, who relocated the woodpeckers to a nearby military base. The only mitigation required of the Red Oak Timber Company was to spend $8,800 (roughly the value of the timber on five or six of the acres harvested) to install and monitor several artificial nesting cavities in a nearby national forest, where by law the Forest Service is supposed to be managing for endangered species already! One can imagine the possible cumulative effects of many such HCPs in a region. We suggest that such abuses of Section 10 are more likely for small, single-species plans than for regional, habitat-based plans that involve many species and many participants and are open to public review.

Single-Species Conservation Plans
Most of the early HCPs were developed for single species. They addressed immediate constraints on land development imposed by ESA prohibitions with apparently little attention to the broader or long-term socioeconomic or biological impacts of the plan and with little attempt to provide the economic or biological certainty potentially offered by habitat-based (multispecies or community-level) planning. The original case study in San Bruno focused on the mission blue butterfly *(Plebejus icarioides missionensis)*, although the plan incidentally benefitted several other protected species and resulted in one species, the Callipe silverspot *(Speyeria callipe callipe)*, not being listed under the ESA. Shortly afterward, other plans were developed for the Coachella Valley fringe-toed lizard *(Uma inornata)*, the Stephens' kangaroo rat *(Dipodomys stephensi)*, the elderberry long-horned beetle *(Desmocerus californicus dimorphus)*, and others.

A common thread of all of these efforts was their largely single-species and small-area focus. In some cases, such as San Bruno Mountain, this focus was effective in resolving the initial conflict, and no other action was required for additional species or habitats (although this plan has been subject to some controversy during implementation). In other cases, such as the Coachella Valley, the original HCP was for a single species but the focus was ecological—that is, the geomorphological processes that produced "blow-sand" and sustained habitat for the lizard. Nevertheless, in this case participants have had to go back to the planning table in order to consider a suite of additional species and habitats that are at risk in the Coachella Valley. As we will discuss, focusing on single species often has proved to be economically troublesome for participants, has been indefensible biologically, and has not solved the problem of additional species needing attention at some later date. But lessons have been learned from these experiences. The limitations of early HCPs initiated an evolution that today has resulted in biologically and politically complex planning processes for multiple species and even entire natural communities and ecosystems.

We hasten to add that for some species, particularly narrow endemics or any others with very small populations, a single-species conservation plan not only is appropriate and desirable but may be the only short-term option. If a species is found only on a small area of private land that is about to be developed, the only alternatives for conserving the species are acquisition of the land by a public or private conservation organization—which is often not possible because of the expense—or development of an HCP to protect the species while developing at least some of the land. In these cases better consideration of the biological requirements of the species concerned can only result in better plans.

In other cases actions must be taken immediately to save a particular species from extinction, and time is not available to consider other species that may deserve attention or to develop a complex multispecies or ecosystem approach. However, expeditious actions on behalf of single species should not preclude ecosystem-level actions at some later date. One would not expect to achieve successful protection for the black-footed ferret *(Mustela nigripes)* in the immediate future through an ecosystem-level approach; captive breeding, intensive translocation efforts, and population monitoring are required. Over a longer time period, however, recovery of the "prairie dog ecosystem" on which the ferret and many other species depend—and which has declined by some 98 percent as a result of government-sponsored poisoning campaigns (Miller et al. 1994)—will be necessary for full biological recovery of the ferret. By the same token, a plant species found on only a few acres, although possible to protect through a multiple species or ecosystem plan, might be short-changed in such a broadscale analysis. The Garrett's mint *(Dicerandra christmanii)*, for example, is found on a small parcel of land in Highlands County, Florida, and was probably never more widespread. All that can be done now to protect this species is to set aside that piece of land and closely monitor the population. Over the long term, however, of what biological value is Garrett's mint separated from the Florida scrub ecosystem in which it evolved? An ecosystem is made of processes, entities, and interactions. When a patch of habitat is too small to support a full community, some entities (other species) are missing and interactions and processes are disrupted. The long-term effects of such disruptions are not easily predictable.

Multispecies and Ecosystem Planning
The conservation of many species together in their natural habitat seems the most logical way to maintain biodiversity. In a sense the emphasis in recent years on individual species conservation is contrary to one of the basic principles of the science of ecology: species exist as functional components of ecosystems. A decline or increase in the population of one species will have effects that ripple through the system in ways that are usually incalculable. Similarly, changes in ecological processes such as disturbance regimes or hydrology will affect many species and communities. The reverse is also true—a change in the biological community from any cause will alter the nature of ecological processes. Most professional conservationists are well aware of the basic principles of ecology—but then why have they pursued the protection of species one by one? It appears that conservationists have pursued a species-by-species strategy largely because they have been fighting a defensive battle. That is, it has been difficult to justify protecting enough habitat to maintain populations of all species or examples of all ecosystems, so conservationists have been forced to rely on emergency arguments: If we don't protect species *x* or species *y* right now, it will be lost forever. Something about the brink of extinction—at least for a charismatic species—seems to wake people up and motivate them to do something. If most species seemed

to be doing well, it made sense to direct scarce financial resources to those species at risk.

Devising conservation plans for species one by one appeared to work well enough for the first few years of the ESA. No one expected recovery to be immediate, even though some species, such as the American alligator *(Alligator mississippiensis)*, bald eagle, peregrine falcon *(Falco peregrinus)*, and brown pelican *(Pelecanus occidentalis)*, showed remarkable progress under the protection afforded by the Act. It was when the list of endangered and threatened species in the United States was approaching 1,000, the backlog of candidate species awaiting protection was expanding ominously into the thousands, and the costs—economic and political—of trying to bring species back from the brink became widely known that arguments for proactive, ecosystem-based conservation began to be taken seriously. J. Michael Scott of the U.S. Fish and Wildlife Service (now with the U.S. Geological Service), who for a while had directed the recovery program for the California condor *(Gymnogyps californianus)*, realized that the millions of dollars spent trying to restore a wild population of the condor could not possibly be generated for every species listed or potentially deserving to be listed under the ESA. Scott, Blair Csuti, and other colleagues suggested that a more efficient and cost-effective approach might be to protect centers of species richness, other biodiversity hotspots, and examples of all vegetation types in a region before the species within these areas and across the broader landscape declined to the point where they required individual attention under the ESA (Scott et al. 1987, 1991a, 1991b, 1993).

Meanwhile, other biologists began to question the exclusive focus on single species in modern conservation. Hutto et al. (1987) charged that "the species approach is too narrow to be used alone as a conservation tool" and specifically criticized use of management indicator species to monitor overall wildlife communities, use of population viability analyses to transform a continuum of persistence probabilities into a simple dichotomy of "viable" and "not viable," conservation evaluations that weigh limited data on single species against better-quantified data on economic value, the disproportionate attention paid to rare and endangered species, and the failure of the species approach to encompass higher-level ecological patterns and processes. Similarly, Noss and Harris (1986) and Noss (1987a) noted that although The Nature Conservancy has advocated a dual approach of a fine filter for species conservation and a coarse filter for conservation of natural communities, until recently most conservation priorities have been based on the locations of rare species, often those that are at the limits of their geographic distribution in the region concerned. Furthermore, the coarse filter, when applied, has focused on small, homogeneous examples of plant communities. Recognizing the ecological processes that occur at a landscape scale, Noss (1987a) urged adoption of an "expanded coarse filter" based on ecological gradients and mosaics and on species that require heterogeneous mixtures of habitats to meet their life-history needs. (Recently, The Nature Conservancy has broadened its approach and is advocating landscape-scale conservation.)

Ironically, or perhaps predictably, the other interests affected by conservation planning have come to a similar conclusion from a different angle. In particular, the private sector is drawn to the conservation planning process and kept at the table by the promise of assusances and certainty—guarantees of streamlined regulations ("one-stop shopping") and a clear picture of future obligations ("no surprises"). These benefits are rarely possible under a single-species planning scenario. Regions of high species richness and high endemism, such as California and Florida, also tend to be the most rapidly urbanizing areas of the country. Addressing one species at a time, without assurances about the nature of future obligations, is neither economically efficient nor practical for the private sector in such situations. Because many of the natural communities in these states are already highly endangered, developing conservation plans for species one at a time, without attention to the larger landscape or ecological processes, only ensures that other species will be listed and need conservation planning relatively soon. Given the high cost of participating in conservation plans, the species-by-species or last-ditch effort approach provides little incentive for most private landowners to cooperate, despite the alternatives. Worse, it may even forestall future opportunities for more effective or proactive conservation action. Sadly, some of the private stakeholders participating in the Stephens' kangaroo rat HCP in Riverside County, California, have apparently become so soured on conservation planning, in part because of the protracted single-species protection efforts there, that they are at present strongly resisting attempts to expand the coverage of the plan to include other listed species—even though it may be in their best, long-term interest.

Early attempts at fulfilling landowners' desire for one-stop shopping focused on multiple species simplistically—that is, two or more species were covered under a single HCP. Most of these multispecies plans were responses to the listing of several species in the same geographic area but not necessarily within the same type of habitat or with any clear mechanism for coordinating conservation actions for species with vastly different habitat requirements. The recently approved HCP for the Balcones Canyonlands outside of Austin, Texas, is a good example of this problem. This plan covers the golden-cheeked warbler *(Dendroica chrysoparia)* and black-capped vireo *(Vireo atricapilla)*, two neotropical migrants with distinct—and occasionally conflicting—habitat requirements, and several listed subterranean invertebrates and amphibians endemic to the Edwards Aquifer system. While this effort managed to achieve many of its biological goals, it was a remarkably difficult and complicated process, and success is not yet assured. In another case, the City and County of San Diego have been involved in a multiple species planning process for more than 90 species, only some of which have overlapping habitat needs. At this writing the process has dragged on for more than seven years, generating great controversy. This plan has been folded into the Natural Community Conservation Planning program described below.

We do not mean to suggest that multispecies plans that involve species associated with different habitats are inherently unworkable or doomed to failure. Indeed, we encourage a true landscape approach—where the mosaic of habitats and species assemblages across a large area are treated as one functional system—in cases where such an approach is feasible and makes ecological sense. There should be no intrinsic barrier to scaling up from different microhabitats or associations within a natural community to different habitats or communities within a landscape. But the conceptual, technical, and practical means for such scaling up must be present at the beginning of the planning process, so that all facets of the plan can be coordinated and integrated throughout the planning process. These kinds of considerations have not been present in most multispecies HCPs or similar plans completed or currently in progress.

The next evolutionary jump—at least in theory—beyond typical multispecies plans came with the development of natural community or ecosystem-level conservation plans. In these plans all the subject species are tied together by a common dependence on or association with a well-defined natural community (although the community may, in fact, be a heterogeneous mix of several related assemblages). The as yet locally unapproved plan for the Florida scrub natural community and 17 listed or candidate species in Brevard County, Florida, is one example (see chapter 5, appendix 5B). Despite the most visible legal incentive being the listed Florida scrub jay, that plan focuses on the overlapping needs of all the species in an imperiled plant community, with the dual goal of providing comprehensive conservation of the community and solid assurances for the private sector. Such plans are entirely consistent with the intent of the ESA, for, as noted above, the first stated goal of the Act is "to provide a means whereby the ecosystems upon which endangered species and threatened species depend may be conserved." It should be remembered, though, that the call for ecosystem conservation in the purposes section of the ESA was never followed up by substantive provisions, and it therefore lacks legal binding authority. But this fact should not dissuade planners from breaking new ground and fulfilling the intent of Congress.

Perhaps the farthest evolutionary step to be formally recognized in conservation law, albeit in somewhat broad and indefinite language, is the California Natural Communities Conservation Planning (NCCP) program. The state legislature in 1991 authorized this program intended to promote cooperation among landowners, developers, urban planners, agencies, environmentalists, local governments, and other stakeholders (Fish and Game Code Section 2800 et. seq., Atwood and Noss 1994). The stated goal of NCCP is to identify and conserve entire natural communities. The pilot project for NCCP is a series of conservation plans for the southern California coastal sage scrub, a community type already depleted by 70–90 percent (Atwood 1993) (see chapter 5, appendix 5A). This is perhaps the largest, most complex conservation planning exercise ever attempted. Plans are being developed for nine separate but geographically adjacent subareas of the southern California region under conservation guidelines developed by an independent Scientific Review Panel appointed by the governor. The FWS has delegated its authority for enforcement and development approval in the process to the state wildlife agency through a special rule under Section 4 of the ESA. Landowners may voluntarily enroll their properties in the program, get state approval for developments that do not conflict with conservation objectives, and ideally avoid the costly delays, permits, and court battles over individual species faced by nonparticipating landowners.

The arguments behind development of NCCP were based on the ten years of experience and lessons learned in conservation planning, as described above. The intent was to use a different scale of planning to address current and future land-use conflicts and provide greater certainty for all participants, conservationists and developers alike, and hopefully for the species that have the most at stake in the plan. However, the conservation intent of the plan has been questioned. The NCCP process was originally promoted politically as a way to avoid listing of species under the ESA and the similar state endangered species law in California (M. Mantell, unpublished testimony, 1991, 1992), but many conservationists were skeptical that the NCCP would make much progress without listings (Atwood and Noss 1994; McCaull 1994). As it turned out, the process needed the legal hook of a listing under the federal ESA to make it functional and to provide "incentives" for landowners to participate. The main initial driving force in the NCCP was the pending listing of the California gnat-catcher *(Polioptila californica californica)*; after much controversy, the species was listed as threatened by the FWS in March 1993 (USFWS 1993). The endangered Stephens' kangaroo rat is also associated, in part, with coastal sage scrub. An additional 33 animal and 61 plant species associated with the community are considered "sensitive"; 53 of them were candidates for federal listing before the C2 category of candidates was eliminated by the FWS (Noss et al. 1995; and see chapter 1). The NCCP mechanism, therefore, should not be construed as a substitute for the prohibitions on taking under the ESA, but instead as a means to address at least some conservation issues before they become conflicts.

Like all conservation planning processes the proof of the merit of NCCP will be in its final results for the biodiversity of the region. It will be many years before these results are clearly evident. Whatever the outcome, lessons learned in the process will refine our ability to address conservation issues effectively in urbanizing landscapes in southern California and elsewhere. Meanwhile, other HCP processes, such as the multispecies HCP currently being designed for the Coachella Valley of California, have been inspired by the NCCP model and not only are considering the needs of many sensitive species but also, in gap analysis fashion (Scott et al. 1993), are seeking to represent viable examples of all natural communities in some kind of reserves (see chapter 5, appendix 5C).

Other Elements of the Endangered Species Program

Habitat conservation plans do not exist independently of other policies and efforts in the federal endangered species program. The listing process, designation of critical habitat, and implementation of recovery plans are all important elements of the overall job of conserving rare and imperiled elements of biodiversity. Ideally, these

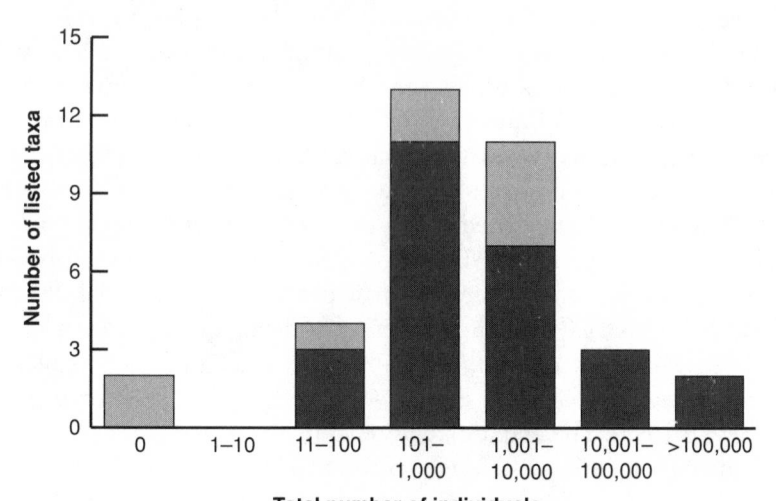

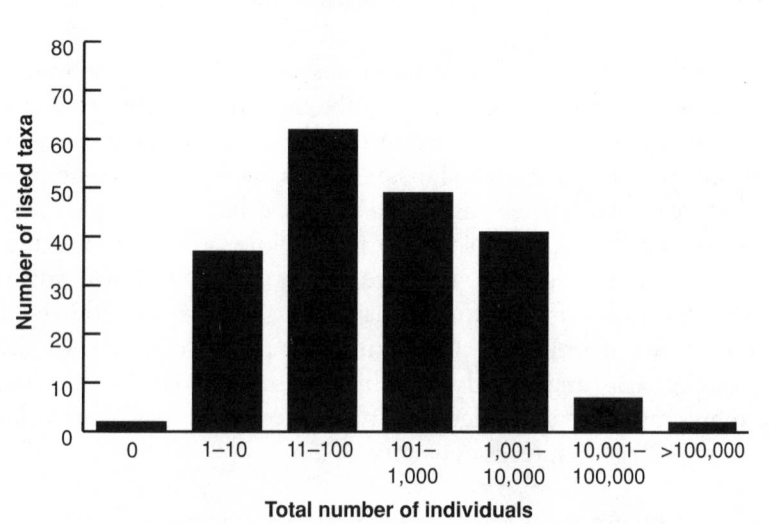

Figure 2.1. Distribution of total population sizes (numbers of individuals) at time of listing for animals and plants. Median size for animals is 999 and for plants 119.5. In the graph of animals, solid bars denote vertebrates and hatched bars invertebrates. From Wilcove et al. (1993).

program components should work together with the result being a coordinated approach to protection. Yet the reality is far from the ideal. For example, Wilcove et al. (1993) studied the listing process and found that by the time most species are listed as threatened or endangered under the ESA, they are in dire need of protection. The median population size at time of listing for animals was 999 individuals, and the median population for plants was 120 (fig. 2.1). The congressionally imposed moratorium on new species listings from March 1995 through April 1996 probably made these figures even more dramatic and limited options for recovery (McMillan 1996). In addition, Congress has been extremely reluctant to provide the agencies with sufficient funds to operate the listing program. Whatever the cause of listing delays, to be effective and allow for flexibility conservation planning must get out in front of the staggeringly slow listing process.

Designation of critical habitat is another element of the endangered species program that has fallen short of expectations. Critical habitat was defined in the 1978 amendments to the ESA, and procedures were specified as to its determination and designation. Bean et al. (1991) pointed out that although the ESA requires that, with few exceptions, critical habitat must be designated upon listing a species, this rarely happens. Between 1979 and 1991, for example, of the more than 400 species added to the list, critical habitats were designated for only 73. Critical habitat has rarely been determined with an explicit conservation planning goal or reserve design in mind and often represents a "best case" for habitat needs that is not achievable. Lack of performance in the critical habitat component of the ESA, however, has had less negative effect on the conservation of imperiled species—particularly on private land—than have deficiencies in the implementation of other elements of the program. Critical habitat has legal effect only when there is some form of federal involvement. The only real effect of designating critical habitat has been to put federal agencies on notice that their activities could be constrained in certain areas; therefore, its impact is relatively inconsequential in most cases (Bean et al. 1991).

The lack of connection between habitat-based conservation planning and other elements of the endangered species program is most serious in the area of recovery. Not only does the number of completed recovery plans lag far behind the number of listed species, but recovery plans rarely are formally coordinated with conservation efforts on private land (where recovery plans legally have no effect). Just as the FWS has few resources to list species, it also lacks funding and staff to develop defensible recovery plans. Houck (1993) cites congressional statistics showing that only 61 percent of listed species had approved recovery plans. By 1996, the FWS's own statistics placed that figure at 54 percent (USFWS 1996). According to Houck's review 23 percent of listed species did not have a recovery plan started, and 30 percent had been listed for more than three years without a recovery plan. Even in cases where recovery plans are adopted they are frequently of poor quality scientifically (Tear et al. 1993, 1995), which makes them of little value for coordination with HCPs. Houck (1993) refers to the approved Mexican gray wolf *(Canis lupus)* recovery plan as an example of a plan containing no estimate of costs for recovery, no timeline, and no specific tasks. Another approved plan for listed cats in the Southwest focuses almost entirely on research and ignores needed conservation actions.

Recovery plans are described in the ESA as the coordinating means by which the law can achieve its ultimate goal of recovering and delisting species. Bean et al. (1991) suggested that recovery plans not only should be coordinated with HCPs, they should identify ways in which HCPs could help further recovery goals. Ideally, HCPs would be nothing less than the private land component of recovery planning. As noted earlier, however, the FWS's draft handbook on HCPs (USFWS 1995) states that, at a minimum, HCPs simply should not preclude recovery. Section 10 of the ESA is not biologically conservative, because it emphasizes the practicality of actions at least as much as their conservation effectiveness. Nevertheless, we feel that no matter what tasks or conservation areas are identified under a recovery plan, if an HCP or series of HCPs undermines this goal or is otherwise inconsistent with it, there will be few opportunities to remedy the problem. Whenever possible, then, the FWS should use what discretion it has to insist on compatibility and integration between HCPs and recovery plans. Sometimes the FWS has recognized this need and promulgated conservation planning guidelines for private lands that complement recovery planning. The Service's Florida Field Office recently commissioned guidelines for the Florida scrub jay (Fitzpatrick et al. 1997) that strengthen the recovery process. But in some cases today HCPs are suspected of interfering with recovery planning for listed species, and at least occasionally (such as for the red-cockaded woodpecker; Peters 1996) they may conflict with recovery objectives. This is a problem that may require a legislative solution.

The above sections have briefly outlined the history, experience, and rationale behind habitat-based conservation planning and revealed some lessons learned. With this backdrop we begin to develop a framework for ecosystem conservation planning that incorporates both single species and natural communities. Before reviewing some basic principles of conservation biology that we believe should underlie conservation planning, in the next chapter we review and attempt to address some of the criticisms that have been levied on habitat conservation planning. As the reader will note, some criticisms are legitimate and point to serious problems that will be difficult to remedy, whereas others reflect some confusion about the purpose and nature of plans or reflect problems that will be relatively easy to mend.

3
CRITICISMS OF SCIENCE
IN HABITAT-BASED
CONSERVATION PLANS

Despite the goal of conservation planning to reach consensus among disparate parties, not everyone is happy about HCPs, NCCPs, and other planning exercises that involve endangered species on private lands. Many local environmentalists have been outspoken in their disregard for the products of the planning process and for the process itself. They are skeptical both of the motives behind habitat conservation planning and of the commitment to long-term implementation on the part of participants (both private parties and agencies). They are concerned about the power and money wielded by special interests in the planning process. They worry about the devolution of authority, fearing that private special interests are able to manipulate the local governmental process more easily than they could the federal regulatory process. They worry about the finality of it all, and that if a plan doesn't work, there will be no recourse for more effective conservation later. And they are uneasy about giving up the chance to review and consider each future development and land-use impact in an entire region. In sum, the decentralized, one-stop shopping approach of landscape-scale planning is unsettling to many environmental participants.

We worry about many of the same things. The spotty (at best) record of HCPs and related conservation plans created over the past decade lends support to many of the fears expressed by environmentalists. Most existing plans are small in area, scope, and species coverage and provide few assurances that they will successfully conserve the biological elements of interest over the long term. The mitigation requirements for many or most HCPs have been grossly deficient biologically (Wilcove et al. 1996; Bingham and Noon 1997). The motives behind habitat-based conservation planning generally have not been altruistic. But should we expect them to be? People everywhere tend to put their own interests before those of others—particularly other species. Inescapably, wealthy special interests wield an extraordinary amount of influence over conservation plans, as they do in society generally.

We propose, however, that well-constructed conservation plans with a solid foundation in science and abundant public input should suffer fewer of these ills. For example, a strong and enforceable implementing agreement provides opportunities for local conservationists to sue for noncompliance or deliberate misimplementation of a plan. The application of the techniques, principles, and analyses recommended in this report would improve the technical aspects of conservation plans significantly. And even given the "no surprises" policy recently promulgated by the Department of Interior, an adequate monitoring program should identify adverse biological trends that require adaptive management (Holling 1978; Walters 1986) to correct. Although changes—especially increases in protected acreage—are often difficult after a plan has been formally approved, adaptive management can and must be accomplished.

HCPs and NCCPs, in particular, were born in political compromise. They reflect both the lack of enforcement success under the ESA and the mounting political forces that seek to weaken the law. As explained by Bean et al. (1991), the ESA's prohibition of habitat destruction that results in harming a listed species has, in practice, "proved difficult to enforce and has at times led to conflicts between economic development and endangered species conservation." To a large extent, conflicts between endangered species conservation—or conservation generally—and economic activity are inevitable. If we, as a society, care about endangered species, we must change the way we conduct much of this activity and the way we address these conflicts in order to reduce risks to species and their habitats. Done well, habitat-based conservation planning is an opportunity to reduce these risks by conducting economic activities in ways that do not further erode the chances of survival of a species or natural community. In some and, we hope, an increasing number of cases the persistence of populations will be enhanced when a conservation plan is implemented (although, as we explained in chapter 2, the law presently does not require this).

The longer one remains involved in conservation planning, the fewer new criticisms are heard. Nearly all criticisms fall into one of two categories: legitimate, fact-based concerns about the planning process or how science is or is not being applied, and criticisms of the HCP mechanism that reflect partisan politics or no-growth and anti-development philosophies. This book is partly an effort to identify, explore, and hopefully correct the problems that lead to the first category of criticisms. In the process we must disregard the latter category, not because it is inherently wrong-headed, but because it has little to offer for the improvement of the conservation planning process.

We address here only those criticisms of conservation planning that directly relate to scientific and technical aspects of habitat-based conservation efforts. We will discuss general issues, not criticisms of specific plans except to illustrate general points. We try to avoid political judgments as much as possible. The listing of a criticism in this chapter does not mean the criticism is necessarily ours or that we think it is legitimate, only that it has emerged from the public discussion of habitat-based conservation planning. To address any of these criticisms adequately, scientists must be involved in the entire planning process (see box 3.1). Finally, the criticisms we respond to below relate specifically to ESA-related plans (e.g., HCPs and NCCPs), not habitat-based conservation planning generally.

CRITICISM: *Habitat conservation plans do not contain a mechanism to ensure scientific input and review.* Participants in habitat conservation plans have a variety of expectations for scientific input and review. The criticism boils down to two distinct issues: the desire of citizens for review of agency planning and permitting decisions by hand-picked, outside scientific experts, and the need to integrate science into planning decisions up front. The first issue is not particularly germane to this report. It stems largely from a strong distrust of regulatory or land-managing agencies and their staffs by the environmental community, based on valid historical experience and the frequent lack of public clarity or disclosure about how permitting and other decisions are made within the agencies. What the citizens often desire is something analogous to a court trial, where a plaintiff provides expert testimony in support of his or her position. We suggest this is neither effective nor particularly valuable in terms of improving conservation planning, because there are generally outside experts on both sides who will gladly support or oppose permitting and planning decisions, often in exchange for a fee. Although this view is cynical, anyone with experience in environmental litigation

Box 3.1. Bringing Scientists into the Process

This book is devoted largely to the technical issues that arise when one attempts to integrate conservation science into habitat-based planning. We discuss both the theory and the application of science to reserve design and land management. Perhaps equally important, though not the main focus of this book, is the process of including conservation *scientists* in planning. Doing this correctly requires understanding, accommodation, and adjustment on the part of all participants in the planning process, including the scientists.

Bringing scientists in to play a fundamental role in the planning process may be troublesome or uncomfortable to some, particularly those in the private sector who worry about scientists becoming advocates against them. Participants on all sides may worry that scientists have their own agenda or are biased toward a particular viewpoint. Agencies can be concerned about "outside" experts influencing their work or decisions, or criticizing agency science. Scientists, too, must learn how to behave in a different setting than they are accustomed to and may enhance perception problems through irresponsible advocacy. Nevertheless, when scientists participate in the conservation planning process constructively and objectively, they can add tremendously to the public credibility and legal defensibility of the plan and at the same time minimize controversy over the outcome. Even more important, they can enhance the ability of a plan to meet its conservation objectives.

Scientists have been integrated into conservation planning in nearly as many ways as there have been plans. Successful efforts have several common elements. First, the independence of individual scientists is important, if not essential. Although it may be difficult to ensure that participating scientists are totally free from preconceived biases toward the plan or the planning process, separating science from the politics and negotiation over other nonscientific factors involved in planning is essential. A structural way to address this is to establish the scientists or group of experts as a separate entity from the stakeholders negotiating the plan and to draw the line clearly for everyone between this group's task of providing expertise and information and the role of the negotiators in advocating specific conservation alternatives or strategies.

Objectivity and independence are important not only to the credibility of the science but also to the ability of scientists to participate constructively in what is often a volatile and emotionally charged process. There is a big difference be-

Box 3.1. Bringing Scientists into the Process 53

tween "scientific opinion" and the personal opinion of someone who happens to be a scientist. Once scientists have crossed the line from providing expertise to vigorously advocating a particular solution (and many have done so), their science can be easily marginalized, and they might as well engage as a member of a stakeholder group (see chapter 4, box 4.1). A scientist's greatest value comes instead from remaining as impartial as possible other than to advocate thoroughly sound science in support of attaining the conservation goals of the plan. There is generally no shortage of conservation advocates willing to provide comments, particularly in regional planning processes. But qualified scientists with expertise in the region and species of concern—and who have the time and willingness to participate in planning—may be hard to find. No substitute exists for the technical expertise and commitment to rigorous, hypothesis-driven methods that scientists can bring to the planning process.

Perhaps even more crucial to bringing scientists into the planning process successfully is timing. As we have noted and will discuss again later in this book, both science and scientists should be woven into the planning process as early as possible, in order to provide the essential theoretical and empirical foundation for planning. For example, the accepted practice of map-based hypothesis testing and reserve design (see chapter 6) is likely to be understood sufficiently for application only by well-trained, experienced scientists. All too frequently scientists are not involved until much later and then are petitioned to pass judgment on the final reserve design or plan, which may have been negotiated to great detail based on many other factors besides science. This role as *post hoc* judge and jury has very little effect other than to increase the public controversy of the plan. When scientific scrutiny comes only as "peer review" of what many participants already consider a final plan, opportunities for improvement are minimal. Although many in the environmental community demand this type of involvement from scientists under the peer review label, it is quite different from the peer review to which research is normally subjected under the scientific publication process. (Most papers submitted to leading scientific journals are rejected, yet we have never seen a plan or project rejected simply because of unfavorable peer review!) This call instead seems to arise from a deep-seated distrust of the public agencies who oversee conservation planning. While peer review of the science that forms the foundation of plans is highly desirable, it should not come at the end, after all negotiation has transpired. Opportunities for significant revision—even rejection—of plans after review should remain. Our experience has shown that front-loading the scientific analysis and review in the planning process, and using critiques to develop a solid reserve design and management program, can essentially eliminate the need for terminal "peer review," except to verify that standards previously endorsed by independent scientists were adhered to.

What types of scientists should be involved in conservation planning? We believe that qualified scientists may come from any of several domains. Academics are perhaps the most frequently perceived as independent, since their ties are generally not-for-profit. But there are many outstanding private sector and agency biologists whose skills can benefit the planning process. Many consultants have considerable expertise in conservation reserve design and management, and can play key roles in planning. In general, there is a fairly small sphere of experts for any given ecosystem, species, or local area who can be tapped to participate in habitat-based planning processes. Maintaining the objectivity and independence of the experts is probably more important than what sector they come from. And of course, the credentials (e.g., publication record, awards) and acceptance of experts within the larger scientific community (e.g., professional societies) are useful measures of competence.

Frequently the question arises as to compensation for scientific expertise. In general, it is easier to attract and retain the focused work of a group of experts if they are well compensated than to rely on voluntary contributions from scientists who have many other responsibilities. Requests for free reviews generally go the bottom of the in-box. Many professional scientists, including academics, must bring in a significant portion of their salary from grants or contracts (including consulting), especially in these days of declining university budgets. If scientists who serve as consultants or reviewers of plans are not compensated, there is a distinct danger that the only scientists available for the task will be from industry (which will pay them for their involvement). There are exceptions to this rule, however. The Brevard County, Florida, HCP for the Florida scrub natural community (see chapter 5, appendix 5B) featured an extremely sophisticated reserve design process and analysis of species conservation needs, developed by an all-volunteer committee of highly qualified experts. But even in this case data collection, collation, and GIS analysis were performed by a paid consulting team. If compensation is carefully managed and publicly disclosed, then it is the preferred method of retaining scientists and assuring they will put the necessary effort into the process in a timely fashion.

Another variable that must be considered in bringing scientists into the planning process is the range of skills required. To some extent this depends on the specific needs of a plan, but there are some areas of expertise crucial to all habitat-based conservation exercises. Individuals with expertise on selected target or indicator species (see chapter 5, box 5.1) are fundamental. Some

BOX 3.1. BRINGING SCIENTISTS INTO THE PROCESS   55

knowledge of other particularly sensitive or specialized species in the planning area must also be available. At a different scale, community and landscape ecology are critical areas of expertise, as is reserve selection and design. At least basic knowledge of ecosystem processes, land management, and restoration ecology is indispensable.

Finally, judicious management of the scientific process is vital to its success. Leadership and facilitation, as well as a manageable size of the scientific group are important. Ideally, there should be a neutral party trusted by all interests involved in the negotiation, and who also has the ability to keep the scientific process moving forward and communicate effectively with the scientists themselves. This facilitator can act as a buffer between the special interests negotiating the plan and the scientists, protecting both their perceived neutrality and their ability to work unfettered by other influences. Generally, the smaller the group of scientists, the greater their ability to work together and resolve differences. This advantage must be balanced with the necessary range of expertise, however. For most plans, a group of ten or fewer is sufficient without being too cumbersome.

To illustrate the above points it is worthwhile to consider case studies and how scientists were integrated. Many scientific processes have been conducted with regard to the northern spotted owl and Pacific Northwest forests, one of which is discussed in box 6.1. In the Brevard County, Florida, example noted above and in appendix 5B, a formally appointed scientific advisory group of five experts, along with a hired academic consulting team, developed three conservation reserve design alternatives that informed the decisions and negotiations of the stakeholder group. The science advisory group contained experts in both species and ecosystem ecology as well as land management and restoration. The process was facilitated by staff from The Nature Conservancy who ensured the independence of the scientists from the stakeholders. The keys to the success of this process (albeit not the final plan; see appendix 5B) was that the scientists maintained their separation from special interests, and the scientific work was conducted prior to the heated negotiations among stakeholders. Before the interest groups began negotiating their preferred conservation option and reserve design, they were given a range of distinct alternatives from which to choose—all of which met the biological criteria for a viable plan.

In many situations it is not possible for scientific teams to engage in lengthy and expensive processes of generating reserve design alternatives. It is still possible, however, for independent scientists to play a central role in ensuring the validity, credibility, and depth of the science that goes into a conservation plan. This can be done by creating a set of specific principles for reserve design, species conservation, and adaptive management on which the plan can be based. This was the approach taken by planners in the southern Orange County subregion of the California Natural Communities Conservation Plan (NCCP), a process that is ongoing at this writing. The NCCP is an attempt to plan for conservation of the critically endangered coastal sage scrub ecosystem over its entire 6,000-square-mile range (see discussions elsewhere). The planning region was divided into several smaller subregions for planning purposes, including the subregion in southern Orange County. The regional program is underlain by a foundation of conservation guidelines developed by an independent review panel made up of conservation biologists. Yet these overall tenets for conservation planning did not offer enough specific guidance for creating plans at the subregional or site level. To overcome this problem planning participants in the southern subregion sought input from a group of independent scientists with expertise ranging from species biology and landscape ecology to restoration and land management. None of the advisors was from a private firm or agency, although the advisors sought input from both those sectors. Most of the advisors were paid as independent scientists by the state of California, while some served voluntarily. They were facilitated by staff from The Nature Conservancy.

The advisors used the general NCCP guidelines to direct their development of a set of detailed, specific principles for reserve design, species conservation, and adaptive management for the subregion. These principles were offered to the landowner's consulting team, which used them to create a set of reserve design alternatives and an adaptive management framework for the plan. Conducting the process in this way accomplished two important objectives. First, the principles created by the independent scientists ensured that appropriate expertise and information were independently brought to bear on the planning process. And second, the principles gave the private interests a common and accepted set of criteria against which to evaluate their subsequent planning efforts.

would be hard pressed to deny it. Furthermore, the issue of scientific "peer" review of completed plans is often promoted as a red herring to hide less noble sounding antidevelopment sentiments. We note that the legitimate need for expert review would be met by proper attention to the second issue.

Aerial view of development encroaching on coastal sage scrub in Orange County, California. Many of the remaining patches of scrub are highly fragmented or restricted to steep slopes, as shown here. Photo by Reed Noss.

Ultimately, successful integration of scientists into the conservation planning process boils down to deciding their most effective position among all the politics, economics, and other factors that also bear on planning decisions. We suggest that front-loading the scientific process, keeping scientists distinct from special interests, and bringing the best information to bear efficiently at all stages of planning, not only make a plan more credible—this approach also allows scientists to feel more comfortable by maintaining their professional integrity and allowing them to do what they do best. They are then more able to affect positively the outcome of a plan.

The second issue, adequate scientific input throughout the planning process, is perhaps the key to resolving the general controversy over the biological aspects of conservation planning. What is needed is a systematic application of conservation principles and techniques (such as those discussed in the following chapters) to optimize the reserve selection, design, and management components of a conservation plan based on the best scientific information, modeling, and thinking available. In cases where data are insufficient to make even minimal reserve design decisions, up-front scientific input may also mean identifying studies that must be completed before planning can progress further. In Brevard County, Florida, for example, four large databases on habitat quality, local land use, species occurrences, and habitat distribution were collected before the reserve design analysis could be conducted. Similarly, a huge amount of information was assembled in order to develop recommendations for management of the northern spotted owl (Thomas et al. 1990 and subsequent reports). But the decision to collect substantial scientific information in both examples was made by the planners in the process and by decision makers at a higher level, who felt that funding such an effort was justified. Without this level of commitment, data may not be sufficient to provide a rigorous foundation for planning decisions, and the plan will be more vulnerable to challenge.

Integrating substantial scientific input is not a legal requirement of the conservation planning mechanism under Section 10 of the ESA; it depends on the choice of the participants. Wildlife agencies have often been reluctant to demand specific studies by permit applicants, preferring to assume a more passive role of accepting plans after they are prepared along with the studies used to substantiate them. Of course, there are exceptions to this rule. California's NCCP program commissioned an independent Scientific Review Panel to craft not only planning guidelines but a short- and long-term research agenda (box 3.2). Local implementation of these guidelines and fulfillment of the research agenda have been troublesome, but nevertheless, they represent a rare conscious and formal attempt to integrate science into the decision-making process.

Many conservation plans have proceeded with minimal scientific input. Most of the small-area HCPs alluded to in the previous chapter fall into this category. Only time will tell how many of these plans succeed at achieving their goals, but it is clear that many are inadequate biologically (Wilcove et al. 1996; Bingham and Noon 1997). As we will discuss in the following chapters, a number of principles and techniques exist that can significantly improve the application of science throughout the planning process. On the other hand, most planning processes do not have a formal means to ensure that the products benefit from the best information and analysis available. Some bills to reauthorize the ESA have contained language requiring increased scientific input to HCPs, an issue that may be settled by the time this book is in print.

Criticism: *Habitat conservation plans do not contain requirements for monitoring.* Scientific monitoring of the implementation of any conservation plan is vital to its long-term success. Most HCPs have a reporting requirement to allow the regulatory agencies to evaluate the progress of the plan, but as Bean et al. (1991) noted, in some cases the oversight agencies do not have the personnel to review the reports. These reports are rarely scientific; most just chronicle the amount of acreage developed or protected and the implementation funds collected. With the nationwide trend toward proliferation of Section 10(a) permits, the ability to monitor the conservation effectiveness of plans is even more crucial, as is the need to provide more biological information in progress reports and scientifically trained agency staffs to review them.

Very few small HCPs contain scientifically based monitoring programs. Some larger plans, such as the draft HCP for the Florida scrub jay in Brevard County, Florida, or the Coachella Valley fringe-toed lizard HCP in southern California, have identified a need for ecological monitoring. In the case of Coachella Valley, the monitoring program has revealed significant new information about the lizard and its habitat that has led to revision and adaptation of the long-term land management strategy. But such monitoring programs appear to be the exception rather than the rule. The recently completed plan for the Balcones Canyonlands area in Texas, for example, contains monitoring and performance criteria for financing but no criteria for monitoring the viability of target species or other ecological phenomena.

Even in cases where monitoring plans exist, the funding to carry them out is meager. The Coachella Valley HCP (not to be confused with the multiple-species HCP now in progress) scientific monitoring plan is conducted entirely by The Nature Conservancy. The recently approved NCCP for the Coastal and Central Areas of Orange County, California, will have an implementation income of roughly $500,000 annually to monitor more than 37,000 acres, including target species populations, exotic species control, and fire prescriptions. While clearly not every acre of the reserves will require active management and associated monitoring every year, it remains to be seen whether $13.50 per acre will be enough to monitor management of the entire system.

Box 3.2. The Natural Community Conservation Plan Research Agenda

The scientists who were charged with developing a conservation plan for the coastal sage scrub and surrounding ecological communities under the southern California NCCP can claim at least one superlative. Never have biologists been charged with planning for so many species across so large a planning area with so little pertinent data. The result was a set of interim conservation guidelines that was long on general principles and short on empirical specifics—guidelines that would benefit greatly from future research. And, while a data shortage is almost a universal characteristic of conservation on private lands, in circumstances where targeted species themselves reside nearly exclusively on such lands, a literal data vacuum is the commonplace.

With a third to one-half of all listed species in this country occurring on nonfederal lands, much of which is private property, real value exists in looking at the elements of the NCCP research agenda. (The NCCP site and species survey guidelines [Scientific Review Panel 1992] are not covered here.) The scientific review panel recognized that most previous studies in the coastal sage scrub had been carried out in response to requirements for governmental environmental review documents associated with specific development proposals. Those studies tended to focus on habitat on the edge of the urban fringe, much of which had subsequently been destroyed. That situation meant that biological information critical to planning was lacking at virtually all spatial scales and at all levels of biological organization, from populations to species to ecological communities. In response, a data gathering exercise was proposed, which included six components:

BIOGEOGRAPHY AND INVENTORY OF COASTAL SAGE SCRUB AND NEIGHBORING PLANT COMMUNITIES. The basic extent and distribution of coastal sage scrub vegetation and its constituent species should be adequately mapped for the region and each planning subregion. Maps of the region should be provided at a scale of 1:100,000, with minimum mapping units of 100 ha (250 acres) and a minimum resolution of 100 m (330 feet). Ideally these maps would be GIS-based. Data layers should include vegetation, urban and agricultural land use, land ownership, topography, climate, distribution of target species, and any available information on species of concern.

For each planning subregion GIS-based (geographic information system) maps (or accurate manually drawn maps based on similar data) should be pro-

Box 3.2. The Natural Community Conservation Plan 61

vided at a scale of 1:24,000 with minimum mapping units of 10 ha (25 acres) and minimum resolution of 30 m (100 feet). Data layers should include those required for the region, as well as specific conditions relevant to the immediate planning area, with emphasis on ground-truthing and verification of data.

PATTERNS OF SPECIES RICHNESS AND COMPOSITION. The intent of the NCCP is to preserve a substantial representation of the biodiversity in the planning area; hence, information on the effect of reserve size and adjoining land uses on plant and animal diversity would help planning decisions. Monitoring of select taxa is necessary to assess the ongoing success of coastal sage scrub conservation efforts. Indicator taxa (such as coastal sage scrub-dependent birds, small mammals, and butterflies) should be employed due to time and funding constraints. Relationships between species richness and composition, and habitat patch area and isolation, should be investigated in sampling programs. These sampling programs will entail surveys for species richness and composition within a carefully selected series of habitat patches in each planning subregion.

DEMOGRAPHY AND POPULATION VIABILITY ANALYSIS. One test of the potential effectiveness of reserve systems is population viability analysis. Time-series data on two target species of birds, the threatened California gnatcatcher and the candidate species, coastal cactus wren, should be gathered in at least half the planning subregions and from representative physical circumstances that span those found across the regional distributions of the species. Data should include territory size, time budgets, reproductive success, survivorship, emigration, and immigration, with separate data obtained both for males and females where possible. Population viability analyses should be carried out for sample populations and metapopulations, and should consider connectivity and environmental effects.

DISPERSAL CHARACTERISTICS AND LANDSCAPE CORRIDOR USE. More information about the dispersal limitations of species that inhabit the NCCP region could help planning for adequate linkages between reserves and could reveal trade-offs between increased reserve size and improved corridors. Dispersal information adequate to allow tests of sensitivity of metapopulation models to connectivity are required. Data from several locations within the planning region during both breeding and nonbreeding seasons should be gathered on the designated target species, mountain lions, coyotes, and representative small mammals and invertebrates.

SURVEYS AND AUTECOLOGICAL STUDIES OF SENSITIVE ANIMALS AND PLANTS. Basic information on the location, abundance, distribution, and natural history of vertebrate and invertebrate candidate species for federal protection and coastal sage scrub–associated plant species of special concern should be gathered from select sites throughout the NCCP region. Each subregional planning exercise should contribute to this regional effort.

GENETIC STUDIES. The maintenance of genetic variation is critical to the long-term viability of species inhabiting coastal sage scrub and neighboring habitat, and will be an important aspect of monitoring populations under a NCCP. Declining genetic variation will be one symptom of inadequate linkages between reserves and can signal a need for changes in reserve management. Baseline data for comparison with future conditions should be gathered at the earliest possible opportunity. Target species and several invertebrates should be sampled from several locations in each subregion. Most genetic data can be obtained with nondestructive sampling techniques in conjunction with other studies that require handling of individual animals.

After three years, application of the NCCP research agenda has been spotty. GIS data layers were developed and many completed by local governments, complementing map products available through California's gap analysis program. Combined, the mapped data are the most extensive and detailed ever assembled for a regional conservation planning effort on private land. A similar success has been a government-funded field effort at 24 locations in three southern California counties. The project represents an unprecedented attempt to characterize variation in species diversity within and between sites dominated by coastal sage scrub. Each site has 10–20 trapping arrays designed to sample reptile, amphibian, bird, and mammal communities, and differentiate between subtle differences in site-specific habitat variables (patch size and shape, edge effects, etc.). Not only is important biogeographic information being accrued, but some information on dispersal of many smaller vertebrates is becoming available, along with new autecological information for many less well studied species.

The intensive demographic studies that could generate population viability analyses have not been sustained. Monitoring efforts with several California gnatcatcher populations that were initiated in the late 1980s lapsed, and ongoing studies have produced fewer than four years of sequential data. No government funding has been made available for demographic studies in the field. And, unfortunately, no genetic studies are under way, although tissue samples gleaned from the vertebrate sampling scheme are being stored for future use.

CRITICISM: *Habitat conservation plans should not provide "certainty," because the biological world is inherently uncertain.* There is much confusion surrounding the issue of certainty in conservation planning. Two facts are clear, however: (1) the biological world is inherently uncertain (see chapter 4), but (2) landowners and other private parties need long-term regulatory predictability and economic assurances to make it worthwhile for them to enter a habitat conservation agreement. Only the most altruistic and economically secure (some would say foolish) private party would agree to foot the bill for every new conservation measure. Uncertainty surrounding the obligations of private parties under conservation laws has been a highly disruptive economic issue. Assurances and a reasonable level of certainty that new and costly requirements for protecting species will not be forced on private parties are what bring them to the bargaining table. The question becomes, how can planners provide sufficient certainty while not undermining long-term conservation goals?

This contentious issue has been raised another notch by the recent "no surprises" policy promulgated by the Department of Interior regarding Section 10(a) permits (unpublished press release). The policy states that once a permit is issued for a habitat conservation plan, no further requirements will be asked of the private participants. The core of the controversy seems to be where and to whom the policy applies. Some observers have taken the policy to mean that the conservation actions undertaken by HCPs are cast in stone and may never be changed. A closer examination reveals that the clear intent of the policy is that in circumstances where additional requirements are needed, the financial burden of those requirements will fall on the public, not the private landowner or business. This appears fair except to those who would place the entire responsibility of conservation on the private sector. On the other hand, if the burden is placed on the public, then the parties acting on behalf of the public—government—must have sufficient funds to accommodate changes in plans.

The title and wording of the no surprises policy seems nonsensical to many biologists and contributes to the controversy. The fact is, Nature is full of surprises (see chapter 4). To submit that there will be no surprises in conservation planning is to suggest a level of scientific predictability that cannot be justified. Adaptive management acknowledges that we cannot predict precisely the outcome of various management actions; therefore, we monitor the situation and make changes in plans and management practices on the basis of new information. Some of the most important information has to do with the effects of implementing a particular reserve design or management practice. For example, if we find that a particular reserve design does not provide enough connections between habitats for individuals for an endangered species of reptile to move safely among sites, then additional corridors (e.g., highway underpasses) may need to be constructed. Or we may learn that a group of rare plants is declining because they are being shaded out by shrubs; then prescribed burning or manual cutting of shrubs may be necessary.

Scientists recognize that we will never be able to predict with great accuracy the outcome of conservation decisions, especially when they are made in the crisis atmosphere surrounding an endangered species that ostensibly stands in the way of economic activity. Besides the inherent stochasticity or chaos of nature, planners rarely if ever have the luxury of sufficient information on species and natural communities to foresee the future. Perhaps the best we can do is optimize our short-term decisions based on the best science available and cross our fingers. Does this mean we should forgo habitat-based conservation planning as a means to achieve conservation? We think not. Rather than debating the merits of providing certainty to private parties, we suggest the debate be shifted to the questions of how and at what level adaptive modification of conservation plans will be funded. At any rate, the reality of working with private landowners dictates that conservation plans must provide reasonable assurances or they will not move forward.

CRITICISM: *Multispecies HCPs and NCCPs undermine the Endangered Species Act by substituting relatively weak protection for rigorous standards of the Act.* The standard of acceptance for a conservation plan under Section 10(a) is to not reduce appreciably the prospects for survival and recovery of a species (USFWS 1995). The NCCP standard is similar but slightly stronger, being defined as "no net loss of habitat value." This is quite a bit different from the general ESA goal of recovery to the point of delisting or the absolute prohibition on take by private parties in Section 9. Despite its appeal to many conservationists, absolute prohibition of impact to listed species is neither a practical nor politically tenable policy. Besides being impossible to enforce, it prohibits any number of otherwise lawful activities that might incidentally harm or kill a listed species, activities that have long been the foundation of economies in many regions, such as agriculture, fishing, timber harvest, and housing development. (Whether these activities are ecologically and economically sustainable, as currently practiced, in a given region is an important but separate issue.)

Holding the line on any future impacts to a protected species is foolhardy for several reasons. The seemingly rigorous standards of the ESA are actually of little or no value in addressing many of the principal threats to species (M. Bean, pers. comm.). The ESA does not, for example, compel the reconnection of already fragmented landscapes, the protection of suitable but unoccupied habitats, the control of invasive exotics or overabundant native opportunists (cowbirds, ravens, gulls, etc.), the augmentation of small populations, or any number of other actions—in addition to prohibiting take—that might be necessary to assure the persistence of a species. Such actions might be prescribed in a recovery plan, but in most cases they are not. Regional habitat-based conservation plans provide the opportunity to take many of these actions—to protect and manage large patches of habitat that might otherwise languish without attention, to identify and protect biological corridors, to slow or stop the encroachment of invasive exotics, and to prevent degradation of landscapes by piecemeal development or recreational activities such as off-road vehicle use. Private habitat-based conservation plans are not a substitute for recovery, but can potentially contribute to it by protecting—at least in part—key source populations or critical landscape linkages on private lands.

Despite their problems, regional habitat-based plans are preferable to the project-by-project, site-by-site review and mitigation that has been the historical status quo and resulted in considerable habitat fragmentation and population declines both of listed and unlisted species. The perspective that such plans undermine the ESA reflects a false sense of security that protection under the Act guarantees survival for imperiled species. Survival and recovery are goals of the Act, but in fact, many listed species—including those with recovery plans—continue to decline because threats from habitat loss and alteration have not been averted (Tear et al. 1993, 1995; Flather et al. 1994; Wilcove et al. 1996).

Some conservationists, on apparently philosophical grounds, oppose any taking of individuals of listed species for any reason. However, an absolute prohibition on taking individual organisms is contrary to the "population thinking" that conservation biologists accept as preferable to an individual-based paradigm. Simply put, individuals come and go; it is the persistence of the population and the species that matters. Even for many endangered species, loss of a few individuals annually as a result of human activity may be fully compensatory—that is, it substitutes for mortality from natural causes and is compensated by recruitment of new offspring to the population. Taking (collecting) individuals of imperiled species for scientific research may sometimes yield information of great utility in preserving the species. On the other hand, for some of the most highly imperiled species whose populations are tiny and in decline, human-caused deaths may be additive to natural causes, and great effort should be taken to reduce all forms of mortality. We suggest that taking always be minimized and that the issue of the appropriate level of taking be considered carefully for each case on the basis of the best available biological information.

CRITICISM: *Habitat conservation plans should not be prepared in the absence of recovery plans.* We agree with this criticism in theory. As we discussed in chapter 2, better integration of all components of the ESA would improve the prospects for listed species and would be generally more efficient and cost effective. But the reality is that fewer than 54 percent of listed species have approved recovery plans (USFWS 1996); if anything, progress on preparing plans is slowing. The FWS simply lacks the resources to prepare, much less implement, recovery plans. In cases where recovery plans exist, many are plagued by lack of information, incomplete or poor application of science, and failure to identify the fundamental measures and milestones of recovery that are needed to translate broad goals into specific actions. There have been some recent exceptions to this trend, such as the recovery plan for the desert tortoise *(Gopherus agassizii)*. Most recovery plans, however, are neither practical nor comprehensive (Houck 1993), and with budgetary constraints there is no reason to expect that they will improve greatly in number or quality within the near future.

It is unreasonable to expect all economic activity or conservation planning to freeze while recovery plans are created or amended to remedy their deficiencies. In fact, habitat is being rapidly lost in many cases. Furthermore, there is increasing legal concern that recovery plans involving private property might constitute regulatory "taking" if landowners are not compensated adequately. This might be true if a recovery plan delineated reserve boundaries that included private lands. This point may be moot, however, because the courts have not yet held recovery plans to be legally binding—for federal agencies, much less private landowners (M. Bean, pers. comm.). In any case, a conservation plan generated by landowners who are committed to a positive outcome is not likely to involve claims of takings from those landowners.

Recovery plans are intended to provide an aggressive strategy or blueprint that would lead to eliminating the threat of extinction for a species and the need for legal protection under the ESA. A common interpretation of recovery by laypersons is that it is intended to lead to a proliferation of the species of concern to the point that it is no longer of conservation concern. This interpretation is incorrect and biologically naive. Many species are so narrowly restricted (e.g., a desert fish confined to a single oasis or an endemic plant found only on one serpentine outcrop) that there could never be a population increase within their native habitat that would make them invulnerable to all threats. The largest population size they have ever attained in their evolutionary history might be so small that a population viability analysis would predict extinction in the near future. More commonly, habitat loss and fragmentation, especially in urbanizing areas, have already progressed to the point where most listed species will forever require some degree of monitoring, active management, and even active protection to ensure their long-term survival. Laissez-faire management (letting Nature take its course) is no longer an option in most regions if we want to maintain biodiversity (Noss and Cooperrider 1994).

Recovery for a great many species, given their degree of endangerment when first listed (e.g., Wilcove et al. 1993), will probably amount to the management and protection necessary simply to allow them to survive over the near to medium term. We often can do no better. We cannot (yet) create new habitat out of parking lots and housing subdivisions, and more such developments are being built every day. Nor can we expect species in fragmented and degraded landscapes to recover on their own to the point where they require no attention. There may, for example, be impenetrable barriers between local populations that are too small to persist without interchange.

The problem still remains, however, of preventing habitat conservation plans from undermining recovery goals when those goals are poorly defined. We believe the answer lies in how habitat conservation plans and the recovery process are coordinated. Bean et al. (1991) offered many useful comments on this issue (see also the following chapters of this book). Recovery plans can be written or amended to identify areas where HCPs would be useful in conserving species. Recovery plans can and should be used to define and catalyze private land conservation efforts. Bean et al. (1991) even suggest that the recovery planning process (such as that developed for the Florida scrub jay separate from the formal recovery plan; Fitzpatrick et al. 1997) could be used to provide guidance for HCPs that do not undermine recovery goals.

Recovery plans can and should make specific recommendations for HCPs and for how HCPs should be evaluated. But without sufficient resources and strategic guidance the FWS will remain incapable of effectively coordinating HCPs and recovery plans. Although we agree that, without coordination, individual HCPs may sometimes undermine recovery, the degree of risk is difficult to quantify. For the moment, because funding often is available for HCPs but does not exist for recovery plans, HCPs have a good chance of protecting habitat on private lands that would otherwise be destroyed. We should also not rule out the possibility that the best habitat-based conservation plans will indeed make recovery plans unnecessary.

CRITICISM: *Reserve systems are designed in HCPs and NCCPs without proper advance scientific study.* Habitat-based conservation plans have generally been developed in a crisis atmosphere with time demands that preclude lengthy scientific analysis. This situation is unlikely to change soon. There is clearly a minimum amount of data necessary to establish reasonable scientific confidence in a reserve design. But satisfying the fundamentals of reserve design does not necessarily require exhaustive, expensive (relatively speaking), and time-consuming studies. Not only do developers want to get the process over with quickly, but more important from a conservation perspective, without expeditious preparation of plans much habitat will be lost as development proceeds unchecked.

Uncertainty in conservation planning is inevitable (see chapter 4). We suggest that two courses of action together will minimize the conservation uncertainty of a proposed reserve system. First, the framework and guidelines offered in this book are intended specifically to strengthen the reserve design process for conservation plans in the absence of complete data. Most of our recommendations, such as application of well-accepted principles of conservation biology to reserve design, require little more than reasonably accurate map-based information about habitat availability and species occurrences and a broad understanding of the ecology of the landscape in question.

The second course of action that can improve the viability of reserve systems is adaptive management (also discussed later). Its relevance here is that given the circumstances surrounding the development of habitat conservation plans, reserve design decisions will almost always be made in the absence of complete data about their consequences. The question "Will the reserve design work?" is not one that can be answered definitively by scientists or anyone else, at least not at the outset. However, an adaptive management program that aggressively monitors and provides feedback to managers so that adjustments can be made in the conservation program will frequently minimize the negative effects of deficiencies of reserve designs. There are conditions under which adaptive management will help improve a conservation plan and others where it will not work. For example, in a situation where two alternative parcels are being considered in a reserve design process, planners may choose the less biologically valuable parcel because of economic considerations. If the more valuable parcel is developed and biological monitoring of the less valuable parcel determines that it is insufficient to support the species of interest—no matter what kind of management is applied—then there is little basis for adaptive management. On the other hand, if development takes a certain number of habitat patches but leaves what data and theory suggest will be sufficient habitat (if restored and managed properly), then with adaptive management different management prescriptions can be tested experimentally, and those that increase the habitat values of affected sites can be continued and those with less positive results can be discarded. With adaptive management, changes in the plan—depending on what options remain—or in specific management practices can, we hope, be made to correct past mistakes.

With some exceptions most conservation plans to date appear to fall short of both remedies. They contain neither a good adaptive management and monitoring program nor the targeted scientific analysis that identifies reserve design or management alternatives and the consequences of those choices. We are encouraged to observe a greater awareness of this issue recently both among conservation planners and wildlife agencies. We hope future conservation plans will apply these principles more broadly.

CRITICISM: *We don't know anything about the species of concern. How can we effectively plan for their survival?* Habitat-based conservation plans are frequently created for species about which we know little other than the information used to list the species (for unlisted species covered by a plan, we usually do not even have that). Their life history requirements, distributions, population sizes, and responses to development, conservation, and management actions may be known only in the vaguest of terms. It is unrealistic (and will be unlikely to stand up in court) to halt all planning until extensive autecological and distribution data can be collected or until detailed population viability analyses have been constructed for all species of concern in a planning area. It is just as unrealistic to complete a conservation plan in the absence of data about how vulnerable species are likely to fare under plan alternatives. What can we do?

For many HCPs the problem of planning in ignorance appears to be a legitimate criticism because planners made little attempt to gather the necessary data, even from the literature. In a few cases planners have tried to do something to reduce ignorance and uncertainty. For example, the NCCP program for southern California coastal sage scrub lacked fundamental autecological and distributional information about the species of concern and instituted a process to address the issue (CDFG 1993). An independent panel of scientific advisors (which included two of us: Murphy as chair and Noss as member) developed a set of survey, research, and conservation guidelines to address the interim period while regional plans were constructed. The survey and research guidelines included detailed protocols for site surveys and target species censuses, and identified specific conservation-biology questions that needed to be addressed before plans could be approved (see box 3.2). The interim conservation guidelines included, among other things, a 5 percent threshold of additional habitat loss. This was the maximum amount of loss that the scientists estimated could occur within the planning area while not appreciably reducing the prospects for survival or recovery of the natural community or species—provided the losses were restricted to patches of low to moderate conservation value (a decision tree was provided to make such determinations). The interim guidelines were intended to relieve some of the immediate pressure for development while some critical baseline studies were conducted for long-term planning.

The survey and research agenda proposed by the scientific panel for

NCCP (box 3.2) was intended to supplement existing information and provide sufficient data to create long-term conservation plans for the natural community. Several studies were immediately undertaken by academic researchers and have yielded valuable information. However, the scientific survey and research program has not met all the goals envisioned by the scientific panel, in part because of a lack of sufficient funding and in part because the federal and state agencies involved were unable to get requests for proposals out and contracts settled in a timely manner. It appears that government "red tape" can produce considerable time lags in initiating studies for conservation planning. In the case of the coastal sage scrub NCCP, a further disappointment is that few sites have been surveyed in accordance with the recommendations of the panel; thus, basic data on distributions and abundances of many species and information on community and ecosystem ecology are still lacking in many areas. Meanwhile, the conservation planning process has proceeded in southern California in earnest.

While the application of the conservation guidelines in the coastal sage scrub NCCP at the local level has been problematic and controversial, it is clear that the approach of identifying needed information and a process for obtaining this information is the only way to remedy data deficiencies. Taking the next steps of funding and contracting surveys and research is apparently where plans are most likely to fall short. We stress that these problems must be remedied if conservation plans are to be scientifically defensible.

CRITICISM: *Habitat conservation plans fail to deal with large-scale biological uncertainty (climate change, periodic disease or epidemics, hurricanes, catastrophic fires).* This issue is not specific to habitat-based conservation planning—it is common to all planning—but it is raised each time a reserve system is put into place. No matter what approach is used to address endangered species, from strict enforcement of Section 9 prohibitions to regional conservation planning, there are many issues where planners are simply unable to make accurate predictions. These issues are beyond the control of planners, yet they must somehow be accounted for in planning. For example, a severe drought occurred in 1996 in the region of the Balcones Canyonlands HCP. Water levels in the southern Edwards aquifer are below critical for the rare fish and amphibian species there. Despite the best efforts of those involved in the plan, protecting the springs and cave entrances and identifying water-level needs have not eliminated the threat of a regionwide water shortage to the species' survival.

Insect infestations in the Pacific Northwest, uncontrolled fires in southern California, and even such seemingly nebulous effects as global climate change can undermine the viability of a reserve system. We cannot realistically expect habitat conservation plans to effectively mitigate global warming. To label them a failure for not doing so is pointless. But to implement a reserve design that has a high probability of failing if projected warming occurs would be irresponsible. The conservation planning process can and should anticipate these issues analytically in population modeling (even when formal population viability analyses are not possible) and reserve design, to reduce the potential for them to undermine the intended functions of reserves. The consequences of a catastrophic fire or epidemic disease, for example, could be minimized by protecting several semi-isolated patches or subpopulations of target species. It is unlikely that a single fire would affect several reserves if they are spatially isolated or if corridor connections could be temporally disconnected by firebreaks. In a similar way, diseases might be contained within single reserves. Generally, multiple, reasonably large reserves spread throughout the range of a natural community or species are less likely to be affected by stochastic events. Reserves connected by dispersal corridors and/or incorporating broad environmental gradients and high habitat heterogeneity are likely to provide for shifting species distributions with climate change (Noss and Cooperrider 1994). The fundamental principle here is for conservation planners constantly to be aware of how their plan fits into the larger ecological context on landscape, regional, and global scales, and over both the short and the long term.

4

PRINCIPLES FOR HABITAT-BASED CONSERVATION

Conservation-related science has developed a degree of public respect and political power that is unprecedented. We see no reason why this influence should not increase along with the continued, rapid growth of the field of conservation biology. A 1993 survey for Defenders of Wildlife conducted by Peter Hart and Associates showed that the public trusts the word of scientists about conservation issues more than they trust environmental groups or the EPA, and much more than the president, news reporters, business leaders, or members of Congress (Defenders of Wildlife, unpublished). A scientifically credible analysis in conservation planning, as in virtually any field, has considerable legal and (often) legislative influence. Conversely, without a defensible scientific underpinning, a conservation plan can be more easily challenged in court and fail to provide the regulatory predictability and other assurances desired by the private sector. Worse, it may fail to protect the species and ecosystems for which it was designed.

Throughout this book we draw from the science of conservation biology in framing questions, interpreting problems, and making recommendations. Like ecology, conservation biology has been largely a science of case studies (see Shrader-Frechette and McCoy 1993). The case study approach (including pilot projects, demonstration projects, etc.) continues to be one of the more fruitful ways to test ecological theories "on the ground" and, when they are successful, to encourage development of similar projects elsewhere. We see habitat-based conservation plans as case studies. With enough case studies, general patterns emerge that can serve as a basis for development of similar plans elsewhere. Although each species, natural community, site, and conservation problem is unique, some principles or empirical generalizations of conservation biology have become clear. These principles provide a starting point for evaluation of conservation options and ultimately for the development of a scientifically credible conservation plan. Case studies and plans should become increasingly robust over time, with more certainty about their outcomes, as the supporting principles are tested and refined. Because data will never be sufficient to eliminate all uncertainty about planning options in a given case, reliance on well-established principles is often the most conservative course.

In this chapter we review some principles of conservation biology that apply to habitat-based conservation planning (see also Wilcove and Murphy 1991 and Noss 1994). These principles, although open to revision, provide the foundation for criteria to evaluate the probability of success of various conservation options. We begin with some philosophical principles, followed by principles that apply to the specific tasks of reserve design and ecosystem conservation. We emphasize that these principles provide only a starting point for conservation planning in specific cases. Most principles have exceptions; often such exceptions "prove the rule." Exceptions to the principles offered below should be looked for in specific cases, because they will probably signal an unusual situation of biological interest, which in turn demands more innovative thinking.

Philosophical Principles

We begin with philosophical principles because every scientific or planning exercise is based on a philosophical foundation and set of values, whether practitioners choose to recognize this basis or not (Kuhn 1970). For example, Western science is based on standards of objectivity and places high value on research that applies a hypothetico-deductive approach; is experimental rather than purely observational; is conducted carefully and objectively with regard to design, methods, and data analysis; is repeatable; and is thoroughly peer-reviewed. Less openly acknowledged, but now being discussed more regularly, are how values and biases of individual scientists influence what they choose to study, their methods of analysis, and how they interpret and publicize their results.

No science is value-free, so it behooves the individual scientist to be honest and admit his or her values and predilections—as well as those intrinsic to the science they study—openly. As Maguire (1996:915) notes, "By being explicitly self-reflective about the underlying values that guide their choice of research subject, method of study, and interpretation of results, conservation scientists can illuminate rather than obscure the connection between values and science." This openness, though still rare among scientists, is evident in the writings of such eminent biologists as E. O. Wilson, Paul Ehrlich, and Michael Soulé, and would appear to be in the public's best interest. The public expects scientists to be honest.

The fear that many scientists have about acknowledging their underlying values probably springs from the belief that the public, decision makers, or peers will ignore or distrust information that appears value laden (Barry and Oelschlaeger 1996). This fear has a legitimate basis, in that findings tainted by the prejudices of individual scientists are properly seen as unreliable. Furthermore, the community of scientists strongly censures—through peer reviews of manuscripts, promotion and tenure committees, job search committees, and the like—those scientists who appear to promote some special interest. Although scrutiny by peers is one of the finest qualities of science, it is ironic that scientists who honestly acknowledge their values and biases may be seen as untrustworthy, whereas those who hide their prejudices behind a facade of pure, objective science may be trusted!

In the case of conservation biology, a community of scientists has acknowledged that biodiversity is valuable and worth saving; thus, conservation biology has been characterized as "mission-oriented" (Soulé and Wilcox 1980; Soulé 1985; Noss and Cooperrider 1994). This does not make conservation biology less rigorous or more likely to display bias than other sciences. Many other applied sciences, such as forestry, engineering, and medicine, are strongly mission oriented. With the goal of conserving biodiversity and ecological integrity acknowledged, it is incumbent upon scientists to use the most rigorous and least biased scientific methods possible, in order to determine the best way of attaining the stated goals in each particular case (see box 4.1). Scrutiny by other conservation biologists is one of the best ways to assure high scientific standards, for as Aldo Leopold once remarked, professionals demand higher performance standards of themselves than do their patrons (Wagner 1996). In an applied science this scrutiny is particularly important because the individual

scientist must shoulder some of the burden of management decisions; moreover, interest groups, bureaucrats, and politicians may try
to influence the outcome of the scientific exercise (Mattson 1996).
Thus, professional standards and peer review can help protect individual scientists from improper persuasion and intimidation.

The philosophical principles reviewed below are meant to serve as a
guide to scientists and planners who must deal with complex ecosystems, insufficient data, uncertainty over results of studies, competing values or hypotheses, and other commonly encountered problems
in real-life conservation work. We do not suggest that consideration of
these principles will lead to easy solution of problems. But we hope
they will provide a backdrop for more conscientious consideration of
planning alternatives.

ECOSYSTEMS NOT ONLY ARE MORE COMPLEX THAN WE THINK, BUT
MORE COMPLEX THAN WE CAN THINK. This statement, attributed to
ecologist Frank Egler, reflects the tremendous complexities scientists
encounter when trying to understand the organisms and systems
they study. Nature at all levels of biological organization—genes,
populations, species, communities, ecosystems, landscapes—encompasses many phenomena that cannot be perceived, measured, or understood using the traditional methods of scientific inquiry. Hence,
uncertainty is high for all results obtained. New developments in science are not likely to break down entirely the perceptual barriers.
Even with the most sophisticated scientific tools imaginable, reductionism and other rational approaches to understanding Nature have
their limits (Capra 1975).

Although ecologists often have a good general understanding of
natural ecosystems, there are always surprises (see below). The proper
response to lack of full knowledge about Nature is humility, that is,
recognizing that we do not and never will have all the answers; thus,
caution and prudence should be exercised when attempting to manage
Nature. This is the approach advocated by Leopold (1949), Ehrenfeld
(1978), and many other ecologists, yet it is absent from some current
proposals for ecosystem management, which arrogantly assume that
science can provide all the information needed to manage ecosystems
wisely (Stanley 1995). A more sensible approach is adaptive management (Holling 1978; Walters 1986), where a degree of ignorance about
the ecosystem and the outcome of management is admitted, but the

system is monitored closely so that adjustments in management can be made when needed.

Admitting our ignorance about many aspects of Nature and being properly humble and cautious is one thing. But failing to move forward with planning and management because we don't know all we would like to know is quite another. The extreme reductionist approach of enumerating each species of plant, animal, fungus, and microbe; tracking the population dynamics of each species; and unraveling their interactions would require a team of experts working for decades in order to understand even the simplest ecosystem. But such extreme reductionism is not needed in conservation planning. Ecosystems, while complex, can be understood well enough to predict their trajectories reasonably well, most of the time (P. Brussard, pers. comm.). A basic understanding of the dominant variables that structure the ecosystem in space and time, coupled with life-history knowledge of those species that play major roles in the ecosystem or are highly sensitive to human activities, will provide enough confidence to move forward with a conservation planning exercise. Then, adaptive management can be used to compensate for mistakes and bring management back on course with the conservation objectives.

NATURE IS FULL OF SURPRISES. One of the striking revelations of modern ecology is that ecological systems are characterized by nonlinear, nonequilibrium, and often chaotic dynamics (Pickett et al. 1992). We can't be sure of what we are seeing today, we can't predict the future very well, and much of what happens will surprise us. The scientific admission of uncertainty and unpredictability stands in stark contrast to the desire of landowners and other private parties for certainty: they want a conservation plan, once signed, to be final and immutable, with no surprises that will cost them money or regulatory headaches at some later time. How can this conflict be resolved?

Conservation planners must be forthright in acknowledging that, of course, there will be surprises—the issue is at whose expense the surprises will occur. As we learn more from monitoring and research, modification of plans will often be necessary to conserve species and ecosystems. This is the whole meaning of adaptive management. We believe, however, that most of the cost of plan modification should be borne by the public at large rather than by the private parties involved in the plan. The health and integrity of the environment—

Box 4.1. Conservation Biologists As Advocates

A longstanding debate among conservation biologists is over what kind and level of advocacy is appropriate for professionals in this field (see the special section on "Conservation biology, values, and advocacy" in *Conservation Biology*, June 1996). Although few biologists today argue that science is value free or that all kinds of advocacy are wrong, a wide range of opinion exists on where to draw the line between responsible and irresponsible advocacy.

The tension is essentially between two core values of the profession: objectivity and public responsibility. If you are objective, you should make your scientific work as free as possible of biases and values (except, of course, the core scientific values of objectivity, detachment, experimental rigor, etc.!). On the other hand, if you are publicly responsible, you should try to make your research relevant to real-world problems and to interpret your findings in a way that you think will help solve those problems. The interpretation phase of science—including making recommendations concerning policy and management options—is encouraged in conservation biology but is inescapably value laden. When one recommends, however cautiously or conservatively, one advocates. The sense of crisis that underlies conservation work intensifies this problem.

In a paper summarizing research on the effects of the Exxon *Valdez* oil spill, biologist John Wiens (1996:595) wrote, "Advocacy can erode the objectivity and rigor of the scientific process. . . . Although there is nothing wrong with predicting a result (all good scientific hypotheses offer predictions), advocacy can bolster such expectations to the degree that contrary evidence is not considered or hypotheses are accepted without supporting evidence." Wiens concluded that "the role of the scientist . . . is to delineate the domain of scientifically supportable statements and to point out instances in which data are used selectively in the pursuit of advocacy."

Although we agree strongly with Wiens that scientists must consider contrary evidence fairly, not accept hypotheses without rigorously obtained supporting evidence, and take care not to cite data selectively, we think he misrepresents the problem of advocacy. What Wiens is criticizing is dishonesty, not advocacy. A dishonest advocate will cite studies selectively and ignore conflicting data in order to support a favored hypothesis. An honest and responsible advocate, on the other hand, will be an advocate for science, for truth, and for Nature. He or she may also choose to interpret scientific findings to policymakers and managers and to make explicit recommendations for how to proceed on the basis of

Box 4.1. Conservation Biologists As Advocates   79

that information. Because they often know the most about the study area, ecosystem, or species concerned, scientists are frequently in the best position to make responsible, well-informed recommendations.

Philosophers of science today generally recognize the positive role of responsible advocacy in science. For example, Shrader-Frechette (1996:913) writes:

> As Aristotle recognized, equal or objective treatment does not mean treating everyone and every position the same, but treating equals the same. If scientists fail to be advocates and if they treat positions of different merit the same, they practice bias. Also, if scientists avoid advocacy, others may make careless value judgments in their work because they know they or their positions are unlikely to be criticized, unlikely to be tested in the marketplace of ideas.

Shrader-Frechette (1996) also commented that if scientists refuse to act as advocates, they can inadvertently serve the status quo and perpetuate "ethical and environmental errors in the status quo." In the realm of conservation we feel scientists have a responsibility to challenge, for example, the widely held view that all "open space" or habitat is of equal value and to debunk such myths as spotted owls nesting on the roofs of K-Marts (this myth was widely circulated by a U.S. congressman) and extinction being perfectly natural and therefore nothing to worry about (which ignores the critical factor of extinction *rate*). In challenging and debunking these myths scientists act simultaneously as advocates for truth and as advocates for conservation. If conservation biologists, in particular, are not advocates for conservation in a general sense, then conservation biology is not distinct from ecology, genetics, botany, zoology, or any other field of biology, and hence risks being seen as superfluous.

Perhaps the best that scientists who work on conservation-related problems can do is openly admit that they value biodiversity and are advocates for conservation—after all, that is why they got into this field. But at the same time, they should not be advocates of particular approaches to solving a conservation problem *until* or *unless* there is an objective scientific basis for taking that position. That is, they must be assiduous advocates for science at all times. Given a biological goal in a conservation plan, it is the job of scientists to determine how to best attain that goal. Throughout the planning process they must strive to set aside preconceived notions of the "best" approach and remain open-minded to all reasonable alternatives. Otherwise, they may consciously or unconsciously ignore data that do not conform to their pet hypothesis. There is no room in conservation biology for "true believers" with a fixed and narrow vision of what must be done. More than enough true believers exist in the environmentalist and developer camps; scientists have a responsibility to help reconcile these disparate positions by advocating strong science and seeking the truth above all.

Because other important values besides conservation exist in society, such as human health and economic prosperity, scientists must also contribute to solutions that seek to harmonize competing values. They must be "team players" in a broad sense. This said, the major responsibility of biologists in the conservation planning process is to the biota. This is a responsibility that comes with knowledge. It is widely agreed that conservation plans should be conservative in the sense of being more willing to err on the side of protecting too much than protecting too little. Once lost, species cannot be replaced. Advocating this position is entirely consistent with scientific objectivity and public responsibility.

which includes the persistence of native species—is a public good. Thus, it benefits the public as well as the environment to provide reasonable assurances to landowners that they will not be held financially accountable for changes in plans. However, it is difficult to envision changes in plans that will cost landowners nothing, at least in time, property values, or other indirect ways. Additional incentives must be found to give private landowners a vested interest in the best possible conservation outcome. Also, in these fiscally conservative, tax-sensitive times we worry about the ability of governments to provide enough funding for plan modifications, especially those that involve additional land acquisition. This issue must be resolved but is outside our scope here.

THE FEWER DATA OR MORE UNCERTAINTY, THE MORE CONSERVATIVE A CONSERVATION PLAN SHOULD BE. Although scientific certainty and an abundance of data are not equivalent, usually with less information on the conservation elements of interest, there is less certainty about how human activities may affect those elements. Some nontrivial level of uncertainty accompanies all planning decisions. For example, predicting whether a project will have a net undesirable effect on a population of an endangered species is not straightforward because information is usually not available on many aspects of the life history of the species (Schemske et al. 1994); without such information it is difficult to predict impacts of projects on particular species. As emphasized in this book, better empirical data, good experimental design, conceptual rigor, abundant and credible references, and more defensible predictive models will help reduce uncertainty and lend support to planning decisions. But the level of information will never be as great as desired, nor will the outputs of models be beyond questioning. A pragmatic view of conservation planning accepts these facts and seeks to minimize the risks to species and ecosystems.

When information on species locations, population sizes and trends, interspecific interactions, responses to disturbance, and other factors is scarce or questionable, the most defensible interim strategy is one that minimizes development and other human disturbance during the time needed to gather the necessary biological information. This is an illustration of the precautionary principle (see box 4.2). For example, when it was discovered that not nearly enough data were available for construction of a long-term conservation plan, the Scientific Review Panel for the coastal sage scrub NCCP in southern California called for an interim plan involving not more than 5 percent loss of habitat in each planning subregion during a period of three to six years over which field inventories and research were to be conducted (CDFG 1993; see box 3.2). Furthermore, habitat losses were proposed to be restricted to patches of scrub having low to moderate conservation value, such as small sites lacking rare species and surrounded by development. If these guidelines were implemented as intended, large patches of coastal sage scrub, well-connected sites, and significant populations of sensitive species would be protected during the interim planning period. Thus, viable options for future conservation would be retained. Because the NCCP process in southern California has not been completed, it remains to be seen how seriously planners took these guidelines in all cases.

THE LESS THE PREDICTED IMPACT OF A PROJECT ON SPECIES OR ECO-SYSTEMS, THE LESS SCIENTIFIC SCRUTINY IS NEEDED. Housing developments, timber sales, or other projects that, according to the best available information and expert opinion, are unlikely to have significant impacts should not require the same level of scientific review as those predicted to have greater impacts. We are not suggesting that smaller projects affecting fewer endangered species require a less rigorous scientific assessment than larger projects affecting more species—the level of rigor, as well as the thoroughness of mitigation, should be uniformly high. We only suggest that the level of scientific scrutiny (including the financial resources put into the scientific assessment and monitoring) should correlate with the magnitude of potential impacts of the project. Predicting the precise level of impact in advance of development will usually not be easy. However, criteria derived from the principles reviewed here should be applied to assess the probable level of impact.

The FWS recognized these issues in their internal handbook on HCPs (USFWS 1995), which included guidelines that created three categories of impact for conservation plans: low-effect, medium-effect, and high-effect (regional). The guidelines define a low-effect HCP as one that affects an ecologically minor portion of a species range, creates conditions that are transitory or of minor invasiveness, or affects few individuals and healthy populations. The FWS also attached a criterion of small size or simple land ownership pattern to low-effect HCPs. Implied by this designation is the recognition that minor impacts require less scientific analysis and review, but not necessarily less rigorous review. Appropriately, processing of low-effect permits by the Service was given a maximum term of three months.

A dilemma results when little scrutiny is given to a conservation plan because it affects only a small area, but in fact the project has significant biological impacts within and beyond that area because the area is of extremely high quality and, for example, holds source populations of the target species. As noted earlier, of the nearly 200 HCPs in process during 1995, most covered very small geographic areas. The vast majority were for single projects of single landowners. With so many small permits it is quite possible for the HCP program to run counter to its original intent of moving away from project-by-project permitting to regional, cooperative planning. In this situation it is even more important for the principles outlined here to be applied in the field, since an accumulation of the effects of multiple small permits with little scientific review could be disastrous. In addition, even with low-effect conservation plans, some level of ecological monitoring should be applied to assess whether the effects of the project are as minor as predicted. Problems caused by cumulative impacts of many small plans can be reduced by sensible anticipation of impacts and incorporation of HCPs into recovery plans.

CONSERVATION GOALS MUST BE CLEARLY STATED AND CORRESPOND TO THE BEST AVAILABLE BIOLOGICAL INFORMATION. Conservation planning exercises sometimes lack clearly stated goals or have goals that are not biologically defensible. Without defensible goals conservation programs are unlikely to succeed. Explicit (though not necessarily quantitative) goals are better than vague goals, and ambitious goals are usually preferable to weak goals. Because few goals are ever fully attained, starting with a compromise may mean ending up with far too little to assure viability, thus violating the precautionary principle (box 4.2).

Recovery plans for many endangered species were written within political constraints that may have biased goal-setting and planning. An assessment of a sample of FWS recovery plans concluded that roughly 28–37 percent of threatened and endangered species are being "managed to extinction" (Tear et al. 1993). Even if the population goals of recovery plans are attained, some 60 percent of listed vertebrate species in the sample would fall below the biological threshold for endangered status suggested by Mace and Lande (1991). Official recovery goals for the Florida panther and grizzly bear have been criticized as being far too low to assure long-term viability of existing populations, much less recovery (Shaffer 1992; Noss and Cooperrider 1994). These examples suggest that politically determined recovery goals may not coincide with biological reality (or as close as we can come to knowing it) and may have to be adjusted upward.

It is imperative that goals correspond to the best available biological information and that the values underlying the goals (see box 4.1) be made explicit (Maguire 1994, 1996). Among the broad goals generally accepted by conservation biologists are (1) representing in protected areas all kinds of ecosystems (natural communities) across their natural range of variation; (2) maintaining or restoring viable populations of all native species in natural patterns of distribution and abundance; (3) sustaining ecological and evolutionary processes within a natural (historic) range of variability; and (4) being adaptable and resilient to a changing environment (Noss 1992; Noss and Cooperrider 1994). To these goals are often added others that deal with human uses of the environment, such as (5) encouraging human uses of the landscape that are compatible with biological goals, while discouraging those that are not.

These ambitious goals would pertain to regional or broader scales,

Box 4.2. Burden of Proof and the Precautionary Principle

Given the complexities and uncertainties that accompany conservation planning, how can we be cautious enough to avoid serious damage to the ecosystem? One direct way would be to put the burden of proof for potential impacts of developments or other activities on those who propose the activities. That is, a development, timber sale, grazing lease, or recreational use is assumed to be destructive unless its compatibility with conservation objectives can be demonstrated convincingly. Such a shift in burden of proof, however, is contrary to tradition in both science and law.

Traditional hypothetico-deductive science puts the burden of proof on those who would claim or predict an effect of a development or other action (Shrader-Frechette and McCoy 1993). In testing hypotheses the statistical significance of a result (the P-value; e.g., $P < .05$) corresponds to the chance of committing a Type I error. A Type I error occurs when a true null hypothesis is mistakenly rejected or an effect is falsely claimed or predicted. An example of Type I error in applied ecology would be claiming that a timber sale led—or predicting that it will lead—to the extinction of a local population of spotted owls, when in fact the owls simply shift their home ranges to adjacent unoccupied habitat.

Conventional research design seeks to minimize the probability of committing Type I errors. In so doing, however, it increases the chances of committing Type II errors or failing to reject a false null hypothesis and failing to perceive a real effect of an activity. The scientific preference for committing Type II rather than Type I errors is congruent with the "innocent until proven guilty" standard in criminal law (Shrader-Frechette and McCoy 1993). It is assumed that acquitting a guilty person is preferable to convicting an innocent person. A conservation analogy would be a preference for approving an environmentally destructive housing development over prohibiting an environmentally benign development.

As the preceding example illustrates, the innocent until proven guilty standard sometime imposes unacceptable risks on society and the environment. In contrast to the prevailing assumption in basic research, in applied sciences, such as medicine, environmental engineering, and conservation biology, Type II errors are now seen by many scholars as more dangerous than Type I errors because they can result in irreversible damage. Examples of such damage include the death of a patient due to side effects of a drug, collapse of a bridge due to a design flaw, or extinction of species when projects claimed to be benign result in

Box 4.2. Burden of Proof and Precautionary Principle   85

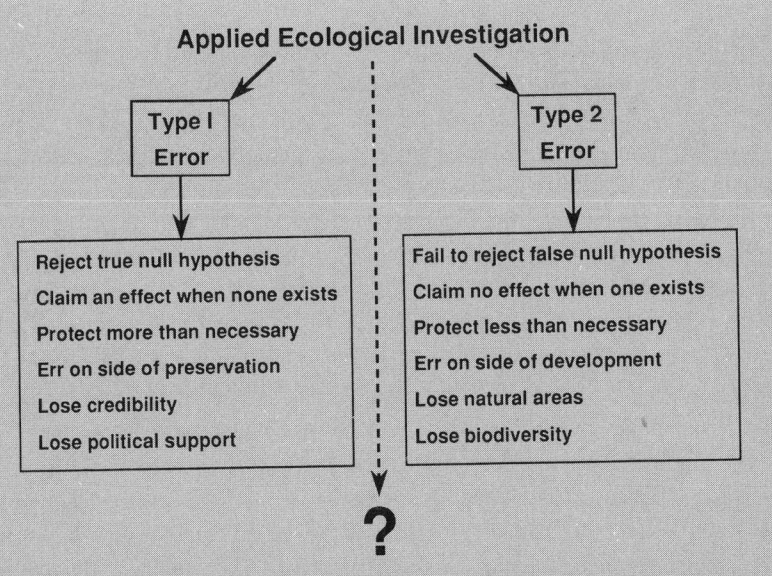

Applied Ecological Investigation

Type I Error	Type 2 Error
Reject true null hypothesis	Fail to reject false null hypothesis
Claim an effect when none exists	Claim no effect when one exists
Protect more than necessary	Protect less than necessary
Err on side of preservation	Err on side of development
Lose credibility	Lose natural areas
Lose political support	Lose biodiversity

?

In basic research, Type I errors are generally seen as more worrisome than Type II errors. In applied sciences, such as conservation biology, both types of errors are problematic, but Type II errors are more dangerous because they can result in irreversible damage to species and ecosystems.

serious habitat degradation (Noss 1986; Peterman 1990a, 1990b; Shrader-Frechette and McCoy 1993). As pointed out by Taylor and Gerrodette (1993):

> Consider a medical test that determines whether a patient has some deadly disease. Physicians are properly less concerned with a false positive (concluding that the patient has the disease when she does not) than with a false negative (concluding that the patient does not have the disease when she does). Conservation biologists deal with the health of species and ecosystems and should be similarly concerned with false negatives.

Although the logic behind a shift in burden of proof is compelling, it will be difficult to accomplish such a shift in environmental law. The legal basis for a shift that would place the burden on private landowners is not established. Significant restrictions on certain uses of private lands to benefit endangered species may constitute a taking for which compensation must be paid (Dwyer et al. 1995a; Houck 1995; and see box 2.2). Although on public lands such a shift may be legally defensible and should be encouraged, on private lands it is unlikely to happen anytime soon (Keiter 1994). Moreover, shifting the burden to private landowners may not be ethically defensible in some cases. For example, the individuals who now own the land on which endangered species reside may often be least responsible for their endangerment, as evidenced by their properties still providing suitable habitat for the species in question. Why should those landowners bear the financial burden of protecting the species, when habitat destruction elsewhere led to the species' decline?

Without taking sides in the debate on shifting the burden of proof from conservationists to those who develop or use land, we believe the philosophy of humility and caution that underlies conservation biology is appropriate and will help reduce the probability of irreversible damage to species and natural areas. In the face of uncertainty we have an obligation to risk erring on the side of increased conservation (see box 4.1). This realization seems to be growing among environmental professionals. As one legal scholar has noted, "The day that is coming does not feature the demise of property rights, as feared as that spectacle may be in quarters as high as the Supreme Court, or even their significant diminution. The day will herald limits to these rights, however, and among them will be the recognition that no one owns the right to extinguish a form of life on earth" (Houck 1995:332).

The precautionary principle, which states that we should act in anticipation of harm in order to prevent it, is gaining increased support in many professions as a way to deal with uncertainty about the potential dangers of activities. Unfortunately, by itself the principle provides no guidance on what precautionary measures should be taken (Costanza 1993); in conservation planning, these measures will have to be determined case by case. Nevertheless, a way must be found to make precaution and prudence standard practice in development, conservation planning, and other risk-prone endeavors of our society. For conservation on private lands this demands that society provide adequate financial incentives to landowners to protect properties that are valuable for conservation or, alternatively, provide sufficient funds for public agencies to acquire those lands.

Until, if ever, the precautionary principle is fully encoded in law and public policy, conservation planners will have to apply the principle in more subtle ways. Recognizing that ecosystems are complex and not entirely predictable, they will design strategies that provide for a high probability of persistence of species and ecosystems, given what is currently known about their requirements for viability. The strategies they develop will be adaptive, that is, able to be modified on the basis of new information derived from ecological research and monitoring. Adaptive management, combined with learning from multiple case studies, will improve knowledge about what precautions should be taken in conservation planning to reduce the potential for biological impoverishment.

and would be easier to apply directly to public than to private lands. Within the area covered by a single conservation plan, such as an HCP or NCCP, and certainly within a single reserve, fully attaining these goals would be next to impossible in most cases. This does not suggest local planners should ignore broad, regional goals; rather, these goals help form the context for reserve design and management decisions at a local scale. The planner or manager must ask, What can I do in this landscape or on this particular piece of land that will best contribute to the broader goals of conserving biodiversity and ecological integrity? It would help local planners tremendously if conservation agencies or organizations would articulate regional goals—even when they lack the force of law—and accompany them by specific objectives and actions required within local landscapes if the broader goals are to be achieved (box 4.3). Further, in the context of adaptive management, a plan should have clear milestones and performance standards that, coupled with an ecological monitoring program, will inform planners, decision makers, and citizens of how well the stated goals are being met.

BIODIVERSITY CONSERVATION MUST BE CONCERNED WITH MANY DIFFERENT SPATIAL AND TEMPORAL SCALES. There is no one best spatial or temporal scale for conservation research or action. Some problems—for instance, incompatible uses or trespass into a local nature reserve—are best handled locally by better visitor education, law enforcement, and facilities management. Other problems—such as designing a timber sale to be compatible with habitat needs of an endangered species that dwells in the area—might be best addressed at a landscape scale (say, 10,000–100,000 ha). Still other tasks, such as designing a reserve network that will maintain long-term viability of large carnivores, must be accomplished at a regional scale measured in millions of hectares (Noss 1992). Similarly, some biological problems, such as a disease affecting a population of an endangered species, must be solved over a period of weeks or months. Others, such as loss of rare alleles from an isolated population, must be managed over decades and centuries.

The trick is finding the best scale for solving each specific problem, then integrating across scales for the overall conservation strategy. As noted above we suggest establishing an explicit hierarchical framework that includes goals, objectives, and specific actions that correspond to different scales in space and time (box 4.3). Long-term, regional goals and strategies guide the establishment of short-term

Box 4.3.

A hierarchical framework of goals, objectives, and management actions for a hypothetical landscape in northern Florida. The examples of objectives and actions are not meant to be comprehensive; many more would be required for an actual management plan. This framework would be appropriate for a regional conservation strategy—for example, as developed by a government agency or private conservation group, and would be unlikely to emerge as part of a private landowner–driven conservation plan. Such frameworks are useful, however, for evaluating the consistency of local conservation plans with broader, regional strategies.

I. Represent in protected areas all kinds of ecosystems (natural communities) across their natural range of variation.

   A. Integrate the Florida Natural Areas Inventory's classification of natural communities with The Nature Conservancy's classification of plant communities for detailed cover mapping of the planning landscape.

      1. Use the Florida Gap Analysis map of vegetation as a starting point, then revise boundaries and increase resolution and discrimination of community types using aerial photographs and extensive ground truthing.

      2. Prepare and digitize maps with a minimum mapping unit of 0.25 ha and display at 1:24,000 scale.

   B. Compare the present vegetation to historical vegetation to determine change in vegetation since settlement.

      1. Construct a map of historic (pre-1800) vegetation for the planning region, using the same natural community/plant community classification as above.

      2. Use General Land Office surveys, notes of early visitors, soil maps, and backward projection from existing vegetation to map historical vegetation.

      3. Compare present vegetation map to historic vegetation map, calculating percent change (% increase or decrease) of each community type and changes in patch size, shape, connectivity, fractal dimension, and other landscape variables since settlement.

   C. Determine community-level protection priorities for the planning region.

BOX 4.3.   89

1. Using digital map of managed (protected) areas from the Florida Natural Areas Inventory, determine level of representation of each current community type in the planning landscape. Compare with information from statewide gap analysis to identify unrepresented and under-represented types at both scales.

2. Using information from analysis of vegetation change (I. B. 3) identify natural communities that have suffered the greatest declines since European settlement in the planning landscape. Compare with information from regional and national summaries of vegetation change (e.g., Noss et al. 1995) to identify "endangered ecosystems."

3. Integrate information from the representation assessment (gap analyses at landscape and state scales) and vegetation change analysis (at landscape and regional scales) to determine protection priorities for community types in the planning landscape.

II. Maintain or restore viable populations of all native species in natural patterns of distribution and abundance.

A. Identify species in the planning landscape that might require individual attention in planning and management.

1. Prepare list of all G3 and above and S2 and above taxa (from The Nature Conservancy's global and state rankings), and others that are listed or candidates for listing under the federal ESA, that have been documented to occur or that might occur (e.g., as predicted from the state gap analysis) in the planning region.

2. Identify other species that are likely to be of high ecological importance or of umbrella or indicator value in the planning landscape (see chapter 5, Box 5.1) and narrow the list of target species to those least likely to be effectively protected by the coarse filter of community representation, most likely to require individual attention, and of highest indicator value.

3. Survey region intensively for species identified above and map occurrences at 1:24,000 scale or larger, using a GPS (global positioning system) for accurate location determination.

4. Search literature for information on life histories, dispersal capacities, etc. of target species and conduct additional, focused research as time and funding permit. Assess population viability of species at a level of detail that available data permit (see chapter 6) and use these assessments to determine additional research priorities.

5. From the above surveys and analyses, identify sites (e.g., habitat patches), critical linkages (e.g., movement corridors), and other landscape features that need protection to prevent losses or declines of species in the planning region.

III. Sustain ecological and evolutionary processes within a natural (historic) range of variability.

A. Build an ecological model of the landscape.

1. Identify and characterize the ecological processes that create and maintain habitat conditions required by target species and natural communities of greatest conservation concern.

2. From historical reconstructions of the planning landscape and results of research conducted elsewhere on similar natural communities, determine the historic and desired range of variability of each of the key natural processes and their interactions.

B. Develop a management strategy for each natural community and for the landscape as a whole, which serves to allow natural processes to operate or mimics them through active management.

IV. Be adaptable and resilient to a changing environment.

A. Develop an ecological monitoring program that is capable of assessing short- and long-term changes in physical conditions (e.g., climate), natural processes, populations of target species, natural communities, and landscape patterns (see chapter 6 and box 6.5 for list of potential ecological indicators).

B. Develop contingency plans for adapting management to ecological change, including long-term changes in climate, sea level (which may inundate parts of the planning region), and biological environment (e.g., invasive exotic species).

V. Encourage human uses of the landscape that are compatible with biological goals, while discouraging those that are not.

A. Identify current human uses of the landscape and determine where they are occurring, with what intensity, and with what biological and ecological effects.

B. Develop a human-use plan for the landscape, including use restrictions (zoning) and a detailed strategy to encourage certain uses (e.g., nondisruptive recreation, environmental education, research) while discouraging others (e.g., off-road vehicle use).

C. Develop an educational strategy and program for the public and for various grade levels, aimed at informing people about their local environment and principles of ecology and conservation biology, and encouraging positive environmental values and behaviors.

objectives and actions, which in turn require more immediate attention. Local management alternatives are evaluated within the context of the regional landscape in terms of habitat availability, dispersal opportunities (connectivity), and maintenance of ecological processes (e.g., wind transport of blow-sand in the Coachella Valley and fire in the coastal sage scrub/chaparral/grassland mosaic) that operate across broad scales.

CONSERVATION BIOLOGY IS INTERDISCIPLINARY, BUT BIOLOGY MUST PROVIDE THE FOUNDATION. Human cultural systems are far more adaptable than biological systems. Thus, although sociological and economic concerns must enter into any conservation planning exercise, the survival requirements of imperiled species cannot be compromised, or, by definition, those species have a high probability of going extinct. Pressure to make economic considerations equal or superior to biological considerations in listing decisions, recovery planning, habitat conservation planning, and other efforts has the potential to undermine these efforts and hasten the slide toward extinction. The biological needs of the target species and the basic requirements for integrity of ecosystems must be determined before any other considerations enter into the planning process. These basic needs and requirements must drive the planning process and not be compromised when economic and social considerations are reconciled in the plans; otherwise, conservation planning will not achieve its primary goals and objectives.

We suggest that biologists and other experts on the species or area in question should not simply "have a seat at the table" in planning exercises or be counted as one of several competing interests. Rather, they must be responsible for determining the requirements for meeting established objectives regarding viability of the species or ecosystems of concern, and should do so independently of the planning negotiations. These objectives, of course—including the choice of time period and the probability of survival that is desirable in viability assessments—are not strictly scientific choices but are social ones that should involve many parties. But once the social choices are made, the biologist's responsibility is to say what it really takes to meet those objectives, not being constrained by social or political pressure. The biologist is also in the best position to evaluate the reliability of existing scientific data and to decide between alternative approaches for meeting biological goals (see box 3.1). Negotiation then can proceed on how best to meet human needs within the biologically determined sideboards. By the same token, biologists have the fundamental obligation to provide this information as free of bias as possible (see box 4.1).

Principles for Species Conservation and Reserve Design

In moving from general, philosophical principles to more specific guidelines for reserve design and management, we should take advantage of what has been learned from experience as well as theoretical considerations. As noted earlier (chapter 1 and box 1.2), effective habitat-based or ecosystem conservation must consider the needs of species whose viability is threatened or that are otherwise of great importance in the ecosystem (see also chapter 5, box 5.1). Most of the research in conservation has been concerned with single species or groups of ecologically similar species, beginning with game management (Leopold 1933) and progressing to nongame and threatened and endangered species conservation over the past three decades. Conservation biology, in particular, has been concerned primarily with the genetic and demographic viability of particular target species.

Diamond (1975) offered a series of rules for "the design of faunal preserves," based on island biogeographic theory (MacArthur and Wilson 1967) and his own knowledge and experience, drawn largely from studies of birds. Other biologists quickly came forward with similar principles. These principles proved to be highly controversial. They were attacked immediately by Simberloff and Abele (1976), stimulating more than a decade of debate over one particular rule, that a single large reserve is preferable to several smaller reserves of equivalent total area. This debate, which came to be known by the acronym SLOSS (single large or several small), finally faded away when most of the participants realized that bigness and multiplicity are both important reserve design criteria (Soulé and Simberloff 1986) and that real-life decisions about reserve design will seldom involve scenarios so simplistic as a single large or several small potential reserves (Noss and Cooperrider 1994).

Although several of Diamond's rules have been discarded because they had too many exceptions, others have endured and a few new ones have arisen. We offer below some principles that have withstood the test of time reasonably well. Although all of these principles have exceptions and are in need of further testing and critical evaluation, we believe they provide a good starting point for modeling reserve design and deciding among alternative designs.

SPECIES WELL DISTRIBUTED ACROSS THEIR NATIVE RANGE ARE LESS SUSCEPTIBLE TO EXTINCTION THAN SPECIES CONFINED TO SMALL PORTIONS OF THEIR RANGE. This was the first of five general principles of reserve design for target species offered by the Interagency Scientific Committee for conservation of the northern spotted owl (Thomas et al. 1990; Wilcove and Murphy 1991). The idea here is that a widely distributed species will be unlikely to experience a catastrophe, disturbance, or other negative influence across its entire range at once. Environmental stochasticity will be important in influencing population growth, but the broader the species' distribution, the less correlated that variation will be among sites. For instance, a five-year drought may severely limit reproduction of a bird species across an area comprising several states; but if the species is distributed throughout the western United States, at least some populations will do well those years and may even serve as a source of colonists to depleted areas.

Keeping species well distributed is therefore a sensible conservation goal and corresponds to the well-accepted multiplicity principle, where it is preferable to have many reserves rather than few (Soulé and Simberloff 1986). (This should not, however, be used as a rationale for habitat fragmentation; see below.) The more populations that remain extant across a broad geographic range, the greater the probability that the species as a whole will persist. Additional support for this principle comes from consideration of the important ecological roles that many species play in local communities, and the genetic distinctness of many local populations (see below). The provision of the ESA that allows for listing of local populations, even when the species as a whole is not threatened, is consistent with this principle. We believe that this provision has not been applied in enough cases and that many more local populations deserve listing under the Act.

LARGE BLOCKS OF HABITAT, CONTAINING LARGE POPULATIONS, ARE BETTER THAN SMALL BLOCKS WITH SMALL POPULATIONS. The principle of bigness is another of the universally accepted generalizations of conservation biology (Soulé and Simberloff 1986; Thomas et al. 1990; Wilcove and Murphy 1991; Noss 1992). Species–area relationships provide one argument for bigness. The larger the area, the more habitats and species it will contain (see review in Noss and Csuti 1994). Species–area relationships, however, do not necessarily argue for large size of individual reserves; many cases exist in which a large number of smaller reserves will contain more species than a small number of larger reserves of equivalent total area. Some of the strongest support for bigness of individual reserves—especially when they are effectively isolated from other reserves with similar habitat—comes from considerations of population viability for species with large home ranges and/or low population densities. A larger block of suitable habitat will usually contain a larger population. All else being equal, large populations are less vulnerable than small populations to extinction as a result of deterministic or stochastic factors (see box 4.4 for a general review of the problems with small populations).

For animals the inverse relationship between population size and extinction is extremely well documented (e.g., Soulé 1987). Large animals, especially carnivores (Noss et al. 1996) have the most demanding area requirements. Although the relationship between population size and extinction probability is not as clear for plants as it is for animals (Schemske et al. 1994), smaller populations of habitat-specialist plants on smaller patches can have higher probabilities of extinction (Quintana-Ascencio and Menges 1996). In line with the preceding principle, large blocks of habitat are also less likely to experience a disturbance throughout their area that could extirpate populations. Thus, refugia and recolonization sources that enhance population persistence are more likely to occur within large blocks of habitat than in small blocks (Pickett and Thompson 1978).

How large populations must be to have a reasonably high probability of persistence for a long period of time is a complex and controversial topic that we will not discuss in detail here. The old rule of thumb that a genetically effective population of about 50 (which usually corresponds to a census population of about 100 to 200 individuals for mammals and birds) is necessary for short-term persistence, and a genetically effective population of about 500 (or a census population of around 1,000 to 2,000) is necessary for long-term survival and evolutionary potential (Franklin 1980; Soulé 1980), may be an underestimate. Even a small probability of catastrophes greatly increases estimates of minimum viable population size and argues for larger populations, as well as multiplicity (Mangel and Tier 1994).

Box 4.4. Threats to Small Populations   95

Box 4.4. Threats to Small Populations

The science of saving species in Nature is really a science of saving the constituent elements of those species—their individual populations. When conservation planners go to work, it is normally a condition of engagement that most "deterministic" or "systematic" threats to population persistence will be terminated or at least regulated. Those threats include harvesting by humans, the accumulation of toxics, and, most important for most species, habitat losses. When biologists initiate reserve design and management planning exercises, the challenge then is to deal with less predictable, often random threats to populations. These "stochastic" threats are usually classified into three broad groups, each of which can contribute to the demise of populations even in the best-planned reserve systems. Unfortunately for populations at most risk, each of the threats increases in importance with decreasing population size.

Genetic Threats
The very smallest populations are challenged by an inescapable fact—should they stay small and not be subject to immigration, they will lose genetic variation. When they lose genetic variation they lose the ability to adapt, to change in response to inevitable environmental variation. Small populations are also more likely to suffer negative effects from inbreeding—that is, mating between relatives. In populations that normally outbreed, inbreeding can decrease heterozygosity (a measure of genetic variation), thus reducing fitness. This inbreeding depression can be manifested in increased offspring mortality and/or decreased body size, fecundity, and longevity (see Lande and Barrowclough 1987). In concert these responses can dramatically reduce the likelihood of population persistence.

Demographic Threats
Just as a population can be threatened by the deleterious consequences of stochastic genetic events, it also can be affected by stochastic demographic events. Unpredictable changes in patterns of births and deaths can cause fluctuations in population size, thus immediately increasing the vulnerability of a population to extinction. Random variation in the sex of offspring, in the average age of first reproduction, in the temporal distribution of offspring over

the lifetime of an individual, or in individual longevity can affect a population's reproductive success and survival rate, therefore potentially reducing its viability.

As with genetic threats, the risks posed by stochastic demographic events increase as population size decreases. The chance occurrence of, for instance, an all-male generation would be much more likely in a small population than it would be in a larger population. However, the size of a population is not the only determinant of its susceptibility to stochastic demographic events; traits such as life history characteristics or reproductive strategies can be just as important. To continue the example, a virtually single-sex generation in a species with nonoverlapping generations quite obviously would be disastrous, while it would not necessarily be so in one with overlapping generations (although, of course, such a birth event is likely only in very small populations). Understanding an organism's reproductive strategy and other life history characteristics is essential to determining its vulnerability to stochastic demographic events.

Environmental Threats

The most apparent factors affecting a population's persistence are those related to its environment. Random environmental phenomena include both abiotic and biotic events that can affect survival, habitat availability and distribution, and resource availability and distribution. Random fluctuations in competition, predation, parasitism, and disease may immediately affect a population, as may abiotic, localized catastrophes such as tree-falls, tidal waves, fires, droughts, floods, and volcanoes.

Unlike genetic and demographic threats, both of which become more serious as population size decreases, the threat of random environmental events often may be unrelated to population size (Goodman 1987). A population's susceptibility to stochastic environmental events may have more to do with its distribution than with its size; species persistence may be correlated with the existence of multiple populations protected on reserves that are unlikely to be affected by the same stochastic environmental phenomena.

Based on a review of empirical and theoretical work on spontaneous mutation and its role in population viability, Lande (1995) concluded that the risk of extinction due to the fixation of mildly harmful mutations may be comparable in importance to environmental stochasticity, and that the estimate of an effective population size of 500

This categorization of these threats to populations—deterministic and stochastic, genetic, demographic and environmental—notwithstanding, extinction almost always results from combinations of threats. Environmental changes, for example, may reduce a population to a size at which inbreeding brings reduced fecundity. The smaller population then becomes more vulnerable to random demographic events, and can become caught in a downward spiral of decreasing viability that leads ultimately to its disappearance. The importance of chance events to a given population depends largely on characteristics of the population itself.

Understanding the characteristics of the population that make it vulnerable to, and that allow it to recover from, such events are as important as understanding the threats themselves (see also Gilpin and Soulé 1986). Perhaps the most important of those characteristics is the number of individuals that are capable of contributing to the next generation, the breeding population size, which is a subset of the entire population. In an ideal world all individuals would contribute equally to future generations, sex ratios would be even, and the rate of decrease in heterozygosity by inbreeding or genetic drift would be predictable. Theoreticians suggest that in such an ideal world, a population of 500 breeding individuals is a reasonable first approximation of the minimum effective size, the size necessary to insure against serious genetic problems (Franklin 1980). In a real population, however, this "magic number" is increased, sometimes dramatically, when the breeding population is a subset of the actual population, when the sex ratio is not 1:1, when progeny survival is randomly variable, when mating is nonrandom, and so on. Thus, for most kinds of organisms the effective size of a population, referred to as N_e, is usually just a fraction of the breeding population. (For example, for mammals such as large carnivores, the ratio of N_e to N, the total population, is often about 1:3 or 1:4; Noss et al. 1996). The calculation of N_e can be of use to and should be considered a component of reserve design when populations are very small and/or when the number of breeding individuals involved in *ex situ* propagation efforts is limited.

being needed for long-term persistence is an order of magnitude too low. Deleterious mutations impose a "genetic load" on populations by reducing survival and fertility. In order to maintain normal adaptive potential under Lande's scenario of genetic load, the genetically effective population size should be on the order of 5,000 for long-term persistence (Lande 1995). For many mammals and birds, this estimate would translate to an actual population of 10,000 to 20,000 individuals. Gilligan et al. (1997) have disputed Lande's estimate on the basis of research with *Drosophila* fruit flies, which showed no trend of smaller populations exhibiting greater genetic load and no trend of captive populations having larger loads than wild ones. It is also difficult to accept that populations as large as 10,000 to 20,000 are normally needed for long-term viability, when narrowly endemic species of many taxa and geographically widespread subspecies of vertebrates have almost certainly never had populations this large, yet they have persisted in many cases for at least thousands of years.

Despite these problems there is good reason to believe that Soulé's (1987:176) earlier generalization of "the low thousands" being the smallest number of individuals to assure a 95 percent expectation of persistence, without loss of fitness, for several centuries, probably applies to most species. Thus, many more species may be biologically endangered than are presently listed, and current FWS recovery goals for listed species are likely too low (Mace and Lande 1991; Tear et al. 1993, 1995; Wilcove et al. 1993). We urge that extent of past decline and risk of further reduction have priority over rarity per se in determining a species' risk of extinction (Noss 1991), as is reflected in the revised IUCN Red List criteria (IUCN Species Survival Commission 1994).

We also reiterate Soulé's (1987:181) admonition that "there are no hopeless cases, only people without hope and expensive cases." Although high estimates of viable population size may seem to lend credence to the view that recovering species to full viability is impractical or beyond hope, there are—simply by chance—many exceptions to any rule. Some small populations, especially if they have been small for a long time and have been purged of their genetic load through natural selection, will persist for a long time. Moreover, viability is not necessarily a threshold phenomenon. Incremental gains in population size bring incremental security, even if we can never boost the population into the thousands (D. Wilcove, pers. comm.).

Recovery goals ideally should be based not on theoretical estimates and empirical generalizations but on detailed knowledge of the genetic structure and evolutionary history of each species individually. Lacking such information, a reasonable strategy is to use the best available models or estimates of population viability (see chapter 6) and to set objectives for restoring the population to a level that has an excellent chance of short- and medium-term persistence (say, for 100 years), a defensible probability of long-term persistence (say, 500 years), and is as close as possible to historical (e.g., pre-European settlement) densities within suitable habitat. One should avoid the temptation to adjust population targets downward to reflect the current availability of high-quality habitat, which may be too little to support populations for very long. Rather than constraining options for survival and recovery, conservation plans should include, where possible, a restoration and management component that will increase the amount of high-quality habitat for target species over time.

BLOCKS OF HABITAT CLOSE TOGETHER ARE BETTER THAN BLOCKS FAR APART. In most cases suitable habitat for a species will not occur in a single patch, but will be distributed among a series of patches. The width and nature of the intervening habitat then becomes an important consideration. Many organisms are capable of crossing narrow swaths of unsuitable habitat, such as a trail, a narrow road, or a vacant lot; far fewer are able to successfully traverse a six-lane highway or the City of Los Angeles. On the other hand, quite a few kinds of animals might travel across a golf course, even though the habitat there is not suitable for breeding. In the absence of impenetrable barriers, habitat blocks that are close together will experience more interchange of individuals of a target species than will blocks far apart. If enough interchange occurs between habitat blocks, they are functionally united into a larger population that is less vulnerable to extinction for any number of reasons (figure 4.1).

Determining the nature of the intervening habitat relative to the dispersal behavior of the target species in question is highly desirable. Blocks of habitat that appear close together to human eyes may be completely isolated from the standpoint of organisms that are physically or behaviorally unable to cross the intervening habitat. Conversely, what appears as a barrier to us may be easily crossed by the species in question. Unfortunately, the dispersal behavior of most species has not been studied, so the best that biologists can do in many cases is make inferences based on knowledge of life histories and information from related species.

HABITAT IN CONTIGUOUS BLOCKS IS BETTER THAN FRAGMENTED HABITAT. This rule follows logically from the previous two but also brings in some new considerations. Fragmentation involves a reduction in size and an increase in isolation of habitats. The theory of island biogeography (MacArthur and Wilson 1967) predicts that either

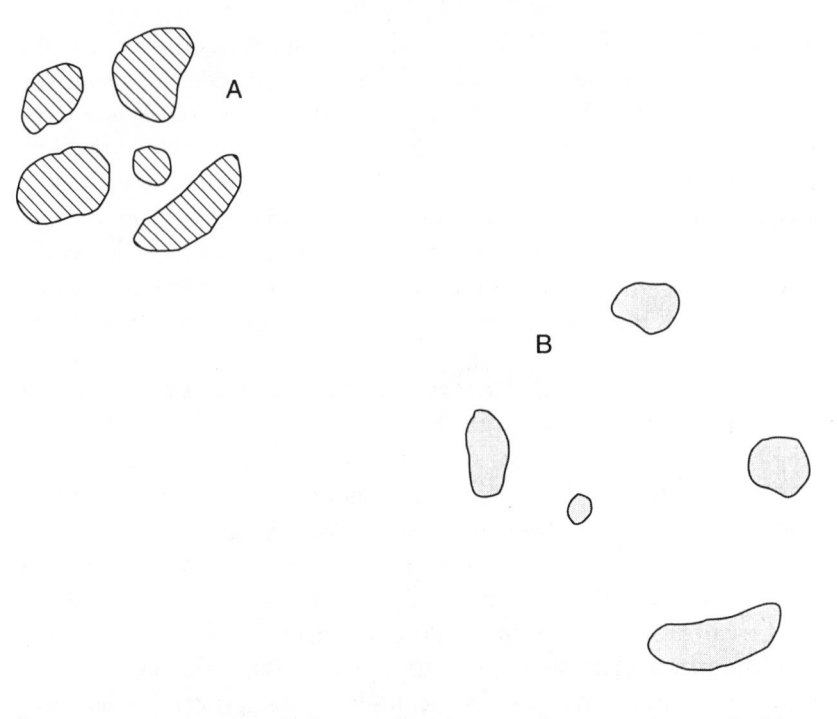

FIGURE 4.1. A group of habitat patches close together (A) is often preferable to habitat patches spaced farther apart (B) because interchange of individuals of the target species among closely spaced patches will create a larger, functional population with a lower chance of extinction. However, careful attention should be paid to the nature of the intervening habitat. If the spaces between habitat patches in (A) are absolute barriers to dispersal, such as highways are for some species, then this pattern is no better than (B). (Figure inspired by a similar drawing in Dunning et al. 1992.)

of these processes will lead to lower species richness due to decreased immigration rates (in the case of isolation) and increased extinction rates (in the case of small size). Thus, a small island far from the mainland is predicted to have the lowest species richness. Looking at a single target species, as is now the fashion in fragmentation studies, a small and isolated habitat patch is expected to have a smaller population and less opportunity for demographic or genetic "rescue" from surrounding populations (Brown and Kodric-Brown 1977). In metapopulation theory (i.e., we define metapopulations generally; see Note from the Authors), an unoccupied patch of suitable habitat isolated by fragmentation is less likely to be colonized or recolonized by a target species (Gilpin and Hanski 1991). If enough connections between suitable habitat patches are severed, the metapopulation as a whole is destabilized and less likely to persist. Extinctions of local populations under these circumstances signal bit-by-bit extinction of the meta-population or the species as a whole (Harrison 1994).

Fragmentation involves more than population effects for single species. Effects at community, ecosystem, and landscape levels are also well documented (Burgess and Sharpe 1981; Noss 1983; Harris 1984; Wilcove et al. 1986; Saunders et al. 1991; Noss and Csuti 1994). Briefly, problems include abiotic and biotic edge effects that reduce the area of secure interior habitat in small habitat patches and often lead to proliferation of weedy and opportunistic species; increased human trespass and disturbance of sensitive habitats and species; and disruption of natural disturbance regimes, hydrology, and other natural processes. The end result of fragmentation is often a landscape that has lost sensitive native species and is dominated by exotics and other weeds. Although species richness at the local or landscape scale is often higher after fragmentation than in the undeveloped condition, many of the added species are organisms that thrive in human-disturbed areas. Higher species richness in human-disturbed landscapes is misleading because it is accompanied by a net loss of sensitive, native species and ultimately by a homogenization of floras and faunas at a broader scale as regions of similar climate begin to converge in species composition (Noss 1983; Mooney 1988).

It should be noted that most studies that have found negative effects of fragmentation have been carried out in highly fragmented landscapes and in regions such as the eastern and midwestern United States, where the matrix is agricultural or urban land and where natural habitat exists only as remnant islands. Many conservation plans under the ESA to date have been carried out in such landscapes. However, studies in less intensively used landscapes where the matrix remains as forest or other natural or seminatural habitat have generally found few negative effects of fragmentation. In such landscapes sharp edges between forest and nonforest are transient because of forest regrowth, and connectivity among patches of similar habitat is high. In central Canada, for example, predation rates on birds' nests, both near edges and in forest interior, were much higher in forest patches in an agricultural landscape than in logged and contiguous forest landscapes (Bayne and Hobson 1997). In the Oregon Coast Range, McGarigal and McComb (1995) found that landscape structure (composition and configuration) explained less than 50 percent of the variation in the abundance of breeding bird species. Species' abundances were generally greater in the more fragmented land-scapes; only the winter wren *(Troglodytes troglodytes)* appeared sen-sitive to fragmentation. However, late seral forests were the matrix in the Oregon study areas, and the species that are probably most sensi-tive to fragmentation (such as the northern spotted owl and marbled murrelet, *Brachyramphus marmoratus*) had sample sizes too small to be included in the analysis (McGarigal and McComb 1995). In addi-tion, sensitivity to fragmentation may be expressed at spatial scales different from those studied.

We can assume that fragmentation is a continuous process and that each species has a characteristic threshold, above which fragmentation becomes a problem. Determining these thresholds (which probably also depend on the nature of the edges and matrix) should be a major research priority, not necessarily of agencies involved in conservation planning, but of the research community generally.

INTERCONNECTED BLOCKS OF HABITAT ARE BETTER THAN ISOLATED BLOCKS. Connectivity—the opposite of fragmentation—has become one of the best-accepted principles of conservation planning. Al-though arguments over the benefits versus costs of particular corridor designs continue (e.g., Noss and Harris 1986; Simberloff and Cox 1987; Noss 1987b; Bennett 1990; Hobbs 1992; Simberloff et al. 1992; and Noss 1993), few conservation biologists disagree that local popu-lations functionally connected by natural movements are less subject to extinction than populations isolated by human activity. Corridors or linkages can be expected to function better when habitat within them resembles that preferred by target species. For example, al-though we do not know exactly the types of habitats all species asso-ciated with old-growth forests will travel through, old forests are likely to provide better linkages for these species than fresh clearcuts.

Some specific guidelines follow from the connectivity principle. They include (a) all else being equal, wide swaths of suitable habitat are better than narrow corridors; (b) corridors longer than normal dis-persal distances for a target species should be sufficiently wide or have enough "stepping-stone" habitat patches to provide for resident indi-vidual home ranges; (c) animals usually follow a path of least resis-tance when moving through a landscape—thus, ridgelines, adjacent parallel slopes, and riparian networks are natural movement corri-dors; (d) planners should base connectivity designs on the needs of species most sensitive to fragmentation (see Noss 1992, 1993, and Noss and Cooperrider 1994).

BLOCKS OF HABITAT THAT ARE ROADLESS OR OTHERWISE INACCESSI-
BLE TO HUMANS ARE BETTER THAN ROADED AND ACCESSIBLE HABITAT
BLOCKS. Noss (1992) added this principle to the preceding five proposed by the Interagency Scientific Committee (Thomas et al. 1990;
Wilcove and Murphy 1991). It applies to species especially sensitive to
human harassment or persecution or sometimes even simple human
presence. Roads and other forms of human access often lead to high
mortality rates for large carnivores, mustelids, amphibians and reptiles (including gopher and desert tortoises), commercially valuable
plants such as cacti, and other species exploited or persecuted by people (Noss and Cooperrider 1994).

Although the ultimate solution to these problems must involve education and change in human values and behaviors, the more immediate need is to restrict access to habitats of sensitive species. For example, land managing agencies often have policies (which may or may
not be enforced) calling for road densities not exceeding 0.5 miles per
square mile in wolf or grizzly bear habitat (Noss 1992). Roadkill is
also a primary source of mortality for many species in regions with
heavy traffic, dirt roads contribute sediments to streams, and roads are
barriers to movement of some small vertebrates and invertebrates.
For these and other reasons (see Noss and Cooperrider 1994), roadless
areas should be protected, roads should be closed whenever possible,
and busy roads should be equipped with underpasses or other wildlife
movement passages. Trails and other avenues of human access also
need to be restricted in many cases or at least routed away from sensitive habitats and restricted to foot use only.

POPULATIONS THAT FLUCTUATE WIDELY ARE MORE VULNERABLE THAN
POPULATIONS THAT ARE MORE STABLE. Just as rarity does not necessarily imply immediate threat of extinction, mean population size is
sometimes a poor indicator of vulnerability. A population with a relatively large mean size but high variance may be more likely to go extinct than a smaller but more stable population (Karr 1982; Pimm et
al. 1988). Species dependent on patchy or unpredictable resources (for
example, requiring ephemeral ponds or rare fruits) often fluctuate
strongly in population size over time and space. Species of arid habitats, such as shrublands and deserts, also commonly experience major
population swings in response to drought (Wiens 1986). Large-bodied
animal species, although more vulnerable to many specific threats,
generally fluctuate less than small-bodied species and therefore can
potentially be viable with smaller populations.

C. D. Thomas (1990) noted that, for species with high levels of population variability, a population geometric mean must be 5,500 individuals to drop below 100 only once every 100 years. He suggested that a minimum viable population size of 1,000 would be adequate for species with normal levels of population fluctuation, and 10,000 should permit medium to long-term persistence of birds and mammals with high levels of fluctuation. What constitutes "normal" and "high" variability is entirely relative and, so far, largely theoretical. Extensive empirical work is needed in this area. An example of relevant research is a study of persistence of herbivorous insects on *Phragmites australis* wetlands. These insects, which show drastic population fluctuations, may require maintenance of habitat patches big enough to support outbreak populations as large as 180,000 adults in order not to decline to extinction during unfavorable periods (Tscharntke 1992).

This principle, like all, will have exceptions, and relationships between body size, population size, population variability, and persistence are extremely complex (Pimm 1993). The addition of deterministic factors, such as overhunting or rapid habitat destruction, may threaten species regardless of their normal amplitude of population fluctuation.

DISJUNCT OR PERIPHERAL POPULATIONS ARE LIKELY TO BE MORE GENETICALLY IMPOVERISHED AND VULNERABLE TO EXTINCTION, BUT ALSO MORE GENETICALLY DISTINCT THAN CENTRAL POPULATIONS. This well-documented pattern is a direct consequence of reduced gene flow to isolated or marginal populations. The pattern presents a dilemma because populations with lower heterozygosity are likely to be less adaptable to future environmental change (Frankel and Soulé 1981) and therefore might be seen as less important to conserve. Marginal populations are also likely to be in suboptimal habitat, of smaller size, and more subject to environmental or demographic stochasticity. Thus, conservation at the species level may be more effective when directed to the central portion of each species' range. On the other hand, disjunct or peripheral populations are likely to have diverged genetically from central populations due to isolation, genetic drift, adaptation to local environments, or some combination of these factors (Lesica and Allendorf 1995). Directional selective pressures can be expected to be intense for these populations, promoting speciation

(Mayr 1963). Thus, conservation of peripheral and disjunct populations promotes future species diversity (Lesica and Allendorf 1995).

Curiously, data on endangered mammals shows that most species have collapsed from the centers of their ranges outward, which also suggests that peripheral populations are worth saving (Lomolino and Channell 1995). Similar findings are emerging for birds (M. Scott, pers. comm.). The explanation for these patterns is not yet clear, but in some cases peripheral populations appear to be refugial; that is, they persist when the central population is annihilated by some catastrophe or epidemic. For example, a tiny peripheral population of the Amargosa vole *(Microtus californicus scirpensis)* near a spring persisted when large flood events eliminated some 99 percent of the vole's occupied habitat (S. Johnson, pers. comm.) Again, the provision of the ESA that allows for listing of distinct populations, even when the species as a whole is not threatened, makes biological sense. Conservation of species across their native ranges is the necessary strategy (see the first principle in this section). Research to determine the genetic distinctness of disjunct and peripheral populations in various regions is recommended, where feasible, as part of the process of target species selection. We hasten to add, however, that conservation of local populations is desirable for their role in local ecosystems and other reasons, regardless of whether the populations qualify as "evolutionarily significant units" or are otherwise of genetic interest.

Principles for Community and Ecosystem Conservation

The increased interest in conservation at the community or ecosystem level carries with it a need for guiding principles analogous to those just discussed for target species. But as noted earlier most of the research effort in conservation biology has been toward the conservation of single species (usually rare ones) or groups of ecologically similar species thought to be in peril (e.g., neotropical migrant birds). Although managers of public and private lands have been managing ecosystems for a long time, their concern likewise has been for particular resources (e.g., timber, livestock forage, game and fish) or public values (e.g., scenery and recreation) rather than for the biodiversity or ecological integrity of the ecosystem as a whole (Noss and

Cooperrider 1994). The new interest in ecosystem management, especially among federal agencies, is somewhat encouraging, but the experiments have not progressed long enough to offer clear lessons, and the motivation behind this interest is suspect (Stanley 1995, but see Christensen et al. 1996). The following are what we characterize as emerging principles for habitat-based conservation of communities and ecosystems.

MAINTAINING VIABLE ECOSYSTEMS IS USUALLY MORE EFFICIENT, ECONOMICAL, AND EFFECTIVE THAN A SPECIES-BY-SPECIES APPROACH. Although many sensitive species require individual attention in order to avoid extinction, focusing on every species individually is impossible. Thousands of species may inhabit any given region, if we include microbes, soil invertebrates, and other poorly known groups. Ecosystem conservation has the advantage of being potentially proactive, by protecting habitats and assemblages before any single species declines to endangerment. If many species associated with an ecosystem are already imperiled, habitat protection and restoration based on their collective needs will be more efficient than single-species recovery (Murphy et al. 1994; Noss et al. 1995).

Numbers of listed and candidate species associated with ecosystem types that have declined as a result of human activities are often high. For example, 27 federally listed species and around 100 Category 1 or 2 candidates (before the FWS eliminated Category 2 in early 1996) are associated with longleaf pine *(Pinus palustris)* and/or wiregrass (*Aristida* sp.) communities in the Southeast (Noss et al. 1995). Conserving and restoring longleaf pine–wiregrass communities across their original range—which would require a strategy far broader than HCPs but is consistent with the recovery objectives of the ESA—would seem more efficient than preparing 27 + individual recovery plans and potentially hundreds or even thousands of HCPs. This is not to suggest that none of those species would need individual attention if an ecosystem-level plan were in place. Several, at least, undoubtedly would need intensive, individualized management until their populations increase to a level where an ecosystem plan would meet their needs. Other species would be useful as ecological indicators and umbrellas (see chapter 5, box 5.1) and in defining the particular species associations of greatest concern. But we suggest the needs of species can be met best when they are considered together within an ecosystem recovery plan that includes attention to the processes that sustain the overall community.

BIODIVERSITY IS NOT DISTRIBUTED RANDOMLY OR UNIFORMLY ACROSS THE LANDSCAPE. IN ESTABLISHING PROTECTION PRIORITIES, CONSIDER "HOTSPOTS." Hotspots are areas of concentrated conservation value. At continental (between-regional) scales, hotspots include centers of endemism and regions of high native species richness for various taxa (see chapter 6, box 6.6). Within regions, hotspots include concentrated occurrences of rare species, habitats of unusually high quality for target species, watersheds of high value for fisheries, and other critical areas. Hotspots of some kind or another (or at least "warm spots") occur in virtually every landscape. Multispecies or ecosystem conservation plans should include such hotspots in reserve designs, whenever possible, but taking care to eliminate exotics from species counts and to exclude sites where species richness is artificially high due to human disturbance favoring weedy species.

Map overlays that display multiple conservation criteria, as in the GIS (geographic information system) maps of gap analysis, can show the locations of hotspots at several scales (Scott et al. 1993). A major argument in favor of the hotspot approach is that large numbers of species or individuals can be protected in a relatively small area. This seductive idea may not be entirely accurate, in part because centers of species richness for different taxonomic groups often do not overlap (Prendergast et al. 1993; Saetersdal et al. 1993; Noss and Cooperrider 1994). That is, at a continental scale, selecting a richness or endemism hotspot for amphibians may do a poor job of protecting reptiles. Similarly, the habitats richest in amphibians within a landscape (e.g., wetlands) may be different from those richest in reptiles. Also, hotspots often may be biogeographic ecotones where species richness is high because many species from adjacent regions are at the edges of their ranges. Although these peripheral populations may be of genetic value (see above), they are also likely to be in marginal habitats, such that focusing protection on these ecotones would provide less security and hope for long-term persistence. For these reasons endemism is generally a better criterion for hotspot identification than species richness. The goal would be to represent all endemic species in protected areas, with highest priority given to centers of endemism for each taxonomic group, where the occurrences of many endemic species overlap. By definition, areas with concentrated occurrences of endemic species are irreplaceable.

In regions with high endemism or otherwise localized species occurrences (for example, much of California, Hawai'i, Florida, Texas, and the Southwest), quite a lot of land—often on the order of 25 percent or more of a region—is required to fulfill the well-accepted conservation goal of representing all species in a reserve system; even more may be required to maintain viable populations of those species occurring in low densities (Noss and Cooperrider 1994; Noss 1996b). However, not all land need be in strict protected areas in order to represent species adequately, so long as the habitats of sensitive species are managed in accordance with their needs. On a between-regional scale, hotspot analyses should be supplemented, where possible, by determinations of degree of threat (see Dinerstein et al. 1995; Noss and Peters 1995). Arguably, requirements for HCPs and other conservation plans should be stricter, in terms of area protected, for regions that qualify as continental hotspots (see box 6.6); however, there is not yet a legal basis to support such a policy.

As a final caution we stress that, whenever possible, species and sites of anthropogenic rarity (where species have declined because of human activity) should be distinguished from those of natural rarity. All rare species are not equally vulnerable to human activities. Extent of decline, both of species and natural communities, is generally a better conservation criterion than rarity, especially when coupled with estimates of current and future threat (Noss 1991; Noss et al. 1995; Noss and Peters 1995). We applaud the recently revised IUCN Red List criteria, which emphasize extent of decline and risk of further decline over rarity per se (IUCN Species Survival Commission 1994; see chapter 6, box 6.4).

ECOSYSTEM BOUNDARIES IDEALLY SHOULD BE DETERMINED BY REFERENCE TO ECOLOGY, NOT POLITICS. Ecosystem boundaries are not entirely arbitrary. Topography, geology, soils, and other factors often create discontinuities on the landscape. Ecosystems can be delimited by species ranges, vegetation, watersheds, physiography, or climatic criteria, all of which can be mapped. Boundaries defined on the basis of ecological criteria are often more useful for conservation planning (especially the scientific aspects of that planning) than those defined by conventional political or administrative jurisdictions. The scale and boundaries of the ecosystem should correspond to the management problems at hand. A comprehensive conservation strategy must consider multiple geographic scales, for example from local sites to ecoregions.

Despite the logic of natural boundaries, practical considerations will often demand that conservation plans follow political (e.g., state, county) boundaries. To reconcile this problem, we recommend that planners follow natural boundaries (e.g., ecoregions) within states, counties, and other jurisdictions and coordinate with neighboring jurisdictions in cases where the natural region of concern crosses political boundaries.

BECAUSE CONSERVATION VALUE VARIES ACROSS A LANDSCAPE, ZON-ING IS A USEFUL APPROACH TO LAND-USE PLANNING AND RESERVE NETWORK DESIGN. Some advocates of ecosystem management favor a "landscape without lines" paradigm, where human activities are spread throughout a landscape but more gently than in the past. This approach is not likely to offer sufficient protection to biodiversity hotspots and areas or species especially sensitive to human disturbances. A concentric zoning model with protection increasing inward and intensity of human use increasing outward from sites of high conservation value or sensitivity is a logical approach to conservation planning (Harris 1984; Noss and Harris 1986; Noss 1987c). Such an approach allows for multiple uses and values across the landscape as a whole, but shields sensitive sites from intensive land uses and potential threats.

Buffer zones are among the best-accepted elements of conservation design (e.g., UNESCO's Man and the Biosphere program; UNESCO 1974; and Hough 1988) but are controversial politically and not necessarily useful in all situations. In some cases it may be preferable to have larger reserves than to put the additional area into buffer zones. Furthermore, poorly managed buffer zones may be population sinks for sensitive species or sources for the invasion of exotics or other opportunistic species into reserves. Nevertheless, it is reasonably well documented (e.g., for songbirds, Wilcove 1985, and Friesen et al. 1995) that reserves surrounded by lands with low-intensity development fare better than reserves surrounded by high-intensity development, such as housing subdivisions. This does not mean we necessarily favor zoning for low-density development over high-density development. If a certain number of residences is to be permitted in a landscape, then high-density, clustered housing would be preferred because more area would be available for conservation. On the other hand, if a certain acreage is to be devoted to development in a landscape, low-density housing would be preferred because more natural habitat would remain around residences and there would be fewer people, house cats, and other agents of disturbance.

In concluding this chapter we emphasize the need to move forward with the scientific information and expertise available. Although we pointed to the uncertainty of scientific knowledge and predictions in many of the principles we reviewed, and to the need for humility and caution, we believe that the ability of scientists to assess effects of alternative management plans and actions through predictive modeling, with feedback from monitoring, is improving all the time. As case studies accumulate, the principles of conservation are becoming better established, and true adaptive management is becoming all the more possible. Ecological and biological science, although it deals with a Nature that defies complete rational understanding, remains the best foundation for conservation planning.

5

CRITERIA FOR ASSESSING THE ADEQUACY OF CONSERVATION PLANS

With the principles reviewed in chapter 4 as background, we can begin to evaluate the adequacy of planning options for conserving species and ecosystems. Although, as we acknowledged earlier, the legal standard under the ESA for compatibility of HCPs with recovery objectives is vague, from a conservation-biology perspective it is indisputable that options that contribute to the recovery of listed and candidate species or the ecosystems upon which they depend should be preferred over those that do not. The objectives of ecosystem conservation and recovery of species are explicit in the goals of the ESA, albeit the means to achieve those goals are nebulous. There is no question that the current status of many species and ecosystems is dire. The ESA lists species as endangered when they are in imminent danger of extinction and as threatened when they are likely to become endangered if current trends continue. Current theory and evidence in conservation biology suggest that, in addition to endangered species, many threatened species and candidates will decline to extinction within decades unless aggressive actions are taken soon to effect their recovery (Mace and Lande 1991; Wilcove et al. 1993). Clearly, habitat-based conservation planning must be reconciled with the ESA's goals of conserving ecosystems and recovering species.

Much of the displeasure that has been voiced by conservationists for Section 10(a) of the ESA stems from experience with some recent HCPs that appear to be merely permits to destroy habitat, inching species incrementally down the road to extinction, rather than well-planned efforts to provide overall benefit to the species. Indeed, the FWS has historically treated HCPs largely as a permitting process, not as a conservation strategy. A conservation plan that leads to a net loss of species viability or habitat value is contrary to well-accepted conservation goals, regardless of how it is interpreted legally. Plans that merely sustain the current, imperiled situation of species are also unacceptable, regardless of whether they are legal. Citizen groups have frequently declared that their investment in helping create conservation plans will not be worth it if such plans fail to contribute in a meaningful way to the survival and recovery of imperiled species. We share these sentiments. But as noted earlier, unresolved issues include who pays for conservation planning and for needed modifications of plans, and what standards should be set for the biological adequacy of plans. In this chapter we offer some thoughts on the question of adequacy.

The Contribution of a Plan to Biological Recovery

Conservation plans should be assessed in relation to biological recovery, not simply legal recovery. However, because of the probabilistic nature and inherent uncertainty of all estimates of viability and extinction risk, what constitutes biological recovery is far from straightforward. We would consider a population biologically recovered when the threats to its existence over the foreseeable future (say, 50 to 100 years) have been appreciably reduced, so that the population has a high probability (say, 95 percent) of persisting and, if possible, growing over that time period. The population goals of many approved FWS recovery plans fall below the population levels at the time of listing, and presumably well below the levels required for biological recovery (Mace and Lande 1991; Tear et al. 1993, 1995). Nevertheless, population size alone is a poor index of recovery because it says nothing about genetic or demographic status, much less about external factors that impinge on the population. There may be cases where a smaller population with threats reduced is more secure than a former, larger but declining or unstable population. Assessments of what constitutes biological recovery should be made for each species individually, using the best available data and expertise. For situations in which no recovery plans have been completed for species affected by a plan, or when knowledge of requirements for biological recovery is highly uncertain, the best a conservation plan can do is be consistent with the principles of conservation biology reviewed in chapter 4.

Whereas HCPs and other conservation plans are not equivalent to recovery plans, legally conservation plans at the very least must not conflict with recovery goals. For example, incidental taking under Section 10(a) of the Act must be both minimized and mitigated. Bean et al. (1991) stated that "what Congress [had] in mind, it seems, is that the mitigating features of a habitat conservation plan *fully offset* the detrimental impact of the authorized incidental taking" (emphasis in original). But going further, the conference committee report accompanying the 1982 amendments to the ESA directed the FWS to "consider the extent to which the conservation plan is likely to *enhance the habitat of the listed species or increase the long-term survivability of the species or its ecosystem*" (emphasis added when cited by Bean et al. 1991). Although the legal basis for contributions of HCPs to recovery has since become more questionable, especially with the FWS's current standard for HCPs as simply not precluding recovery, ideally an HCP should contribute to recovery. In addition, recovery plans should provide explicit guidelines for how HCPs should serve this mission. Recognizing that conservation planning will always take place within a political atmosphere and that compromise on the part of all participants is often necessary, it is still essential that the conservation goals for a plan be assessed in terms of their potential contribution to recovery. Use of the best available scientific information and opinion to make such assessments is also essential.

For all practical purposes many regional conservation plans today are, within the region of concern, the equivalent of the private land component of a recovery plan. Regional plans that take both public and private land into consideration may, in the best cases, substitute for recovery plans if they entirely encompass the range of the listed species. Although we do not necessarily condone the substitution of HCPs for recovery plans, if funds are available for the former but not the latter, we believe they should be put to good use. For this reason alone it is important to ensure that conservation plans do not conflict with recovery goals. With the criterion of recovery potential in mind we can evaluate conservation planning options in terms of the questions posed below.

Habitat Value Increase
Is there a net gain in habitat value for the target species (or a net gain in high-quality acreage of the target ecosystem)? Over what time period? Measures of habitat quality depend on adequate knowledge of a species' life history requirements (e.g., substrate preferences, food items, nesting sites) and comparative data on the demographic

performance (e.g., fertility, survivorship) of the species in sites that differ in quantifiable ways. For some species, especially vertebrates, habitat suitability indices (HSIs) have been developed and tested; for many others little is known about the factors that determine habitat value, and more field studies are needed before even general statements about the relative value of different habitats can be made. Especially needed are studies on the effects of different management treatments and other human land uses on habitat quality.

The distinction between habitat quality and quantity is important. A gain in acreage of poor-quality habitat may not provide much benefit and could be a detriment to recovery if the site is a population sink (i.e., where mortality exceeds reproduction) for the target species. Protection of small areas of high-quality habitat may have greater benefits for the target species than larger areas of poor-quality habitat, because the high-quality habitat may support higher densities (this is not always true; e.g., sometimes nonbreeding individuals congregate densely in inferior habitat), has lower mortality rates, and produces more offspring. Both quality and quantity of habitat must be taken into account, both for target species and for natural communities or ecosystems.

Habitat quality often can be affected dramatically by management (see below), to the extent that the present quality of a habitat may be misleading—what is important is the potential quality. Furthermore, not just the internal habitat quality or *content* is important but so also is the habitat configuration or *context* on a landscape scale (Noss and Harris 1986). For example, a site with excellent habitat structure for a species may still be unsuitable if it is small and isolated from sites of similar habitat, particularly if the surrounding landscape is developed or otherwise degraded and the site suffers from edge effects.

Habitat in Conservation Management
Is there a net gain in suitable habitat that is firmly protected and adequately managed? What is the commitment to and funding for management? Over what time period? Conservation options should be evaluated on the basis of expected increase in net habitat value over specified time periods. In the short term we can expect a net loss of habitat acreage or quality in a landscape undergoing rapid human population growth and development, regardless of whether a conservation plan is in place. Nevertheless, in many cases these losses can be compensated for in the medium and long term if habitat quality is

gradually increased in protected sites through restoration and management. Conversely, a gain in suitable habitat in the short term will not benefit the target species or ecosystem if a lack of protection or management lead to degradation later. For example, many sites require prescribed burning or control of invasive exotic plants, feral animals, livestock, or human visitors in order to maintain habitat quality. The establishment of expansive reserves may give a false sense of security that certain species or communities are being protected if that protection is not followed by perpetual management that maintains or improves habitat quality. Well-considered conservation plans contain performance standards to ensure that implementation produces a net gain in protected and adaptively managed habitat.

Increase in Population Size and Viability

Is there or will there be a net gain in population size and/or persistence time (in 5, 10, 25, 50, 100, 200, 500 years) for the target species? This criterion follows naturally from the previous two. Population viability is directly related to habitat quality in a broad sense (including relationships with other species in the community) and to factors intrinsic to the population, such as genetic variation and reproductive potential. Population viability analyses (PVAs) can be conducted to determine the minimum population size necessary to achieve some specified probability of persistence over a specified time period (Shaffer 1981; Boyce 1992). In reality, there is rarely enough time to gather sufficient data and conduct a thorough PVA during planning stages. There are alternatives to PVA, however, that require less data but can provide extremely useful information on viability and species-specific habitat quality (see chapter 6). Viability assessments—PVAs or otherwise—can be used to predict differences in the probability of persistence under alternative reserve designs or management actions and help guide implementation. Because even simple assessments of population viability can be conducted for relatively few species, the target species in a habitat-based conservation plan must be selected carefully and with a solid ecological justification (see box 5.1).

Rationally, all else being equal, the reserve design and management plan that offers the highest probability of persistence for target species should be selected. However, political or economic constraints may preclude implementation of the biologically optimal alternative. Habitat-based conservation plans, especially those involving significant

Box 5.1. Designation of Target Species (the Fine Filter)

Habitat-based conservation plans, even when they are targeted explicitly at natural communities or ecosystems, must pay serious attention to species (see box 1.2). An ecosystem comprises its species, their interactions with each other and the physical environment, and a suite of biotic and abiotic processes. Of these elements, processes (e.g., photosynthesis, herbivory and predation, decomposition, nutrient and hydrological cycles, disturbance and recovery) are arguably the most fundamental, whereas species are the most sensitive. Over time species come and go, while processes live on. But although it is perfectly natural for species to come and go over geological and evolutionary time (hundreds to thousands of years), the rate at which species are going today is, on a global scale, many times faster than the rate at which they are coming. The present rate of extinction is matched only by the great mass extinction events of the distant past and far outpaces the rate of speciation (Wilson 1992).

If processes are relatively stable and species are relatively vulnerable, then species warrant our concern as conservationists—assuming we value biodiversity. But as discussed throughout this book, the species-by-species approach to conservation has proven inefficient and burdensome. We cannot begin to worry about each of the thousands of species that might inhabit a particular conservation planning region or even about the dozens or hundreds we suspect might be highly sensitive to human activities. For this problem The Nature Conservancy developed the concept of the coarse and fine filters (Noss 1987a; Noss and Cooperrider 1994). With the coarse filter we inventory, protect, and manage viable examples of each of the natural communities in a given region. By so doing we protect the vast majority of species (estimated as perhaps 80–90 percent) without having to worry about them individually. Species that are not well protected by the coarse filter include narrow endemics and wide-ranging animals. These species require individual attention—the fine filter. They are among the species that might be targeted for special consideration in habitat-based conservation planning.

Besides the coarse/fine filter rationale for selecting target species, there are other reasons having to do with the functioning of ecosystems, the role of species in that functioning, and the role of species in determining whether the functioning is, in fact, adequate to keep the system intact. For example, we might know from historical data and experience that fire is important in maintaining a particular grassland or savanna community. That information alone,

BOX 5.1. DESIGNATION OF TARGET SPECIES   117

however, is not sufficient for developing a prescribed burning program that will maintain the composition and structure of the community as it occurred historically. We have to find out what intensity, frequency, seasonality, and spatial pattern of fire is "best" for the community. Usually that will be the regime closest to what the dominant, native species experienced during their evolutionary histories. Because, in most cases, we cannot ascertain the historical regime directly, we must experiment—apply adaptive management—to assess the responses of species to fire. Furthermore, we need to determine those species that, perhaps because of their flammability, play pivotal roles in the ecosystem by promoting a fire regime that other species depend on; for example, longleaf pine seems to be such a species in the Southeastern Coastal Plain (Platt et al. 1988).

Other species are critical for securing and cycling key nutrients through the ecosystem or in determining the composition of the food web through their role as predators (Power et al. 1996). Some species, because of their demanding area requirements or dispersal behaviors, determine the necessary size and configuration of habitat patches in the landscape. Still others are highly sensitive to changes in habitat structure and serve as ecological indicators for that reason. No single species is likely to be an adequate umbrella, in the sense of protecting all other species if its needs are met (Noss and Cooperrider 1994, Noss et al. 1996). Therefore, successful habitat-based conservation planning will depend on identifying a set of species whose spatial, compositional, and functional requirements encompass those of all other species in the region (Lambeck 1997).

We follow Lambeck (1997) in suggesting that the first step in identifying target species (what Lambeck and many others call "focal species") is to identify the processes that threaten the ecosystem in terms of contributing to population declines of native species (but note: not all threats can be foreseen; adaptive management must recognize new threats as they arise). Then, groups of species are identified whose vulnerability is attributable to a common cause, such as loss of area or fragmentation of a particular habitat type, competition from invasive weeds, or an altered fire regime. Next, species in each group are ranked in terms of their vulnerability to those threats. Finally, target species are identified whose requirements for protection or management encompass all others in their group Lambeck 1997). Lambeck deals with a landscape that is already degraded but can potentially be restored. In contrast, conservation planning as discussed in this book usually involves landscapes that will undergo further development, leading to further degradation unless mitigation measures are completely successful. Thus, for habitat-based conservation planning, different scenarios of future landscape change must be considered in selecting target species.

Lambeck (1997) identifies area-limited species, dispersal-limited species,

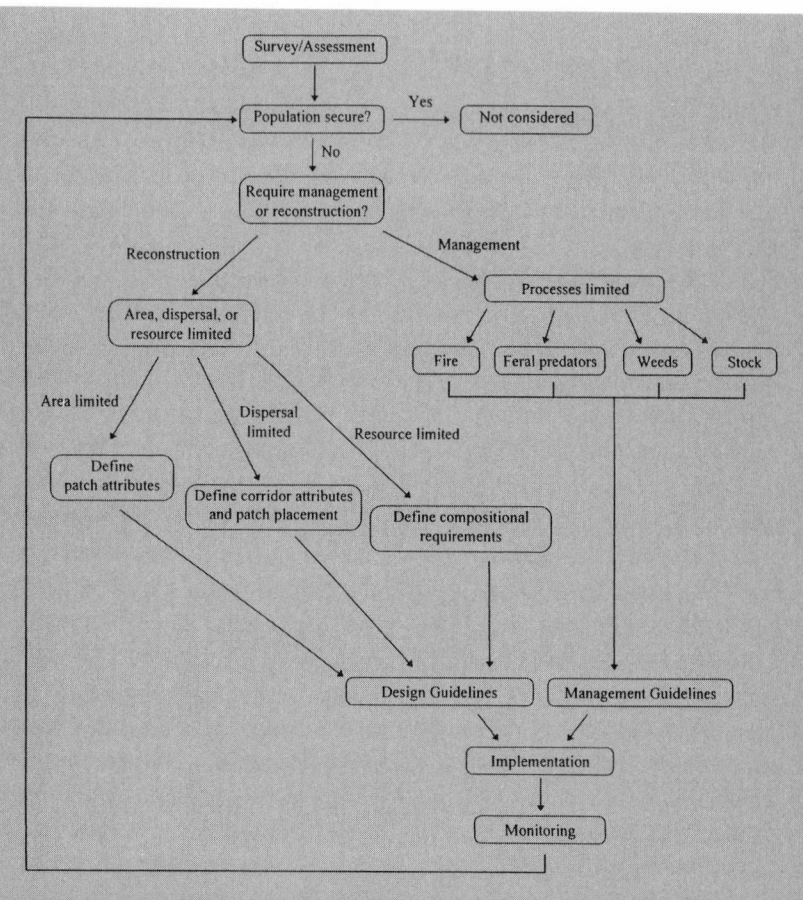

A schematic representation of the procedure proposed by Lambeck (1997) to identify target (focal) species. This example is taken from the Wheatbelt of Western Australia, where habitat loss and fragmentation and resource depletion were identified as the limiting factors. Fire, exotic predators, and weeds are examples of the types of processes that need to be managed. From Lambeck (1997).

resource-limited species, and process-limited species as his vulnerability groups. For each group the focal species are the ones most demanding for the attribute that defines that group. Two or more species might be selected within a group, and a single species may occur in more than one group. For example, the species with the largest home ranges or lowest densities in various habitat types will define the needs of the area-limited group; the species most reluctant to travel through the developed landscape—or which experience the highest mortality trying to do so—define the dispersal-limited group; the species most dependent on frequent fire or periodic flooding define the process-limited group; and so on.

BOX 5.1. DESIGNATION OF TARGET SPECIES   119

When data do not exist to determine with certainty the species with the "most," "highest," "lowest," etc., within these categories, biologists will have to use expert judgment to select species that probably deserve these qualifiers. The needs of these species define the thresholds—in terms of patch size, isolation, fire frequency, level of feral predator control, and so on—that must be exceeded if the full native biota in the region is to be maintained (Lambeck 1997). We acknowledge, however, that it is unrealistic to expect each landscape to contain all the species historically native to it. For example, in many regions lack of sufficient wild habitat, high road density, and high human population density may preclude maintenance or reestablishment of populations of large carnivores in the foreseeable future.

To Lambeck's four groups of focal species we add three final categories: keystone species, narrow endemics, and special cases. Keystone species are defined broadly to include every species "whose impact on its community or ecosystem is large, and disproportionately large relative to its abundance" (Power et al. 1996:609). Some keystone species might be encompassed within one of Lambeck's four groups, but some would not. Keystone species include predators, such as sea otters *(Enhydra lutris)*, that enhance species diversity by holding in check dominant species that would otherwise out-compete others; engineer species, such as beavers *(Castor canadensis)*, gopher tortoises *(Gopherus polyphemus)*, alligators *(Alligator mississippiensis)*, and cavity-excavating birds, which create and maintain habitats used by many other species; key pollinators and seed dispersers; plants that provide resources utilized by many animal species at critical times of the year; and large herbivores that control vegetation structure through their feeding, trampling, and other activities (Noss 1991; Power et al. 1996). For keystone species it is especially important to maintain populations that are close to ecologically optimal, not just minimally viable, because so many other species depend on them. A minimally viable population of a keystone species may be too small or too dispersed to support populations of some obligate dependents, for example the arthropods that are found nowhere else but deep inside the burrows of gopher tortoises (Jackson and Milstrey 1989).

The sixth category of target species, narrow endemics, include those that occur at very few sites (all within an area of, say 50,000 km^2; Terborgh and Winter 1983; Gentry 1986) or occasionally only a single site in the world. If these species are to persist, those few sites where they occur must be protected. The narrower the range and fewer the occurrences, the more stringent the protection required. Finally, the seventh category of species comprises special cases. In this group we include disjunct or peripheral populations that are genetically distinct (and, therefore, may be on the verge of becoming narrow endemic species; see chapter 4), and other species that play strategic roles in a conservation campaign. For example, certain "flagship species" have public appeal that transcends their ecological importance or sensitivity. Many of these species—the giant panda *(Ailuropoda melanoleuca)*, Florida panther *(Puma concolor coryi)*, and wolf *(Canis lupus)*—are the classic "charismatic megavertebrates" and would also fall into the area-limited species category. We can imagine cases, however, where other kinds of species (e.g., distinctive plants like the saguaro cactus *Cereus giganteus*) are highly valued by the public and deserve special attention for the support they help garnish for conservation efforts.

Before finalizing the group of target or focal species we suggest one further screen: the coarse filter. By applying a gap analysis procedure (e.g., Scott et al. 1993) opportunities to represent habitat types that are currently unrepresented or underrepresented in the study region will appear on the vegetation maps. By protecting these areas the needs of some of the target species may be met, along with (presumably) the needs of species in the group they represent. We agree with previous predictions that the vast majority of species—especially those, such as soil microfauna and bacteria, that are little known and difficult to inventory—will be protected by a sensibly applied coarse filter. Those species that remain after the coarse filter screening comprise the list of target species for the planning region. We suggest that these species be subject to population viability assessment of one type or another (see chapter 6) and that their responses to management experiments be monitored rigorously throughout the life of a conservation plan. To summarize, the species to emphasize in conservation planning are:

areas of private land, cannot realistically be expected to provide the best possible conditions for the species or communities concerned, simply because there are other factors at play. On the other hand, a plan that reduces the probability of persistence of a target species population below a reasonable viability threshold should be automatically rejected. One must strive to predict the net change in population viability from the current situation and distinguish it from that offered by other alternatives (including the default option of no conservation plan) while continually refining the predictions over time. The alternative that offers the best prospects for persistence, well above the biological minimum but within politically set constraints, is

BOX 5.1. DESIGNATION OF TARGET SPECIES   121

Area-limited species: Species that require the largest patch sizes to maintain viable populations. These species typically have large home ranges and/or low population densities.

Dispersal-limited species: Species that are limited in their ability to move from patch to patch, or that face a high mortality risk in trying to do so. These species require patches in close proximity to one another, movement corridors, or crossings across barriers such as roads.

Resource-limited species: Species requiring specific resources that are often or at least sometimes in critically short supply. These resources may include nectar sources, fruits, mineral licks, and the like. The number of individuals the region can support is determined by the carrying capacity at the time the critical resource is most limited (Lambeck 1997).

Process-limited species: Species sensitive to the level, rate, spatial characteristics, or timing of some ecological process, such as flooding, fire, wind transport of sediments, grazing, competition with exotics, or predation.

Keystone species: Ecologically pivotal species whose impact on a community or ecosystem is large, and disproportionately large for their abundance.

Narrow endemic species: Species restricted to a small geographic range (e.g., <50,000 km^2 is a commonly used cutoff) and often with very few occurrences within that range.

Special cases: Species important in the planning region that do not fall within one of the above categories. This group includes disjunct or peripheral populations that are genetically distinct, and "flagship species" that promote public support for a conservation plan.

---

generally preferred. Note that biological constraints must be recognized first (see chapter 4), with other constraints considered after the needs of target species and the ecology of the system are well defined.

The implementation term is critical in any comparison of conservation plan alternatives. An option that provides for very high probabilities of persistence over a 10-year period may fail miserably if evaluated over a 100-year time span. Other options may perform poorly over the short term but provide high probabilities of persistence over 200 or 500 years, assuming the population makes it through the short-term crunch. The assumption of persistence through a crunch may or may not be justified. For these reasons the variation in probabilities of persistence over time should be modeled, graphed, and considered in planning decisions whenever possible. We do not place much faith, however, in model projections over long time periods (e.g., 500 years) because it is impossible to predict changes in environment (e.g., climate), human population, land-use trends, political scene, and many other factors over such a span with any accuracy. In most cases it is more useful to be able to predict and compare probabilities of persistence under specific management options over the next 5 to 100 years. It is also desirable that models be spatially explicit; that is, that predicted demographic rates, whenever possible, be associated with particular habitat polygons on a map. Spatially explicit approaches to PVA and other viability assessments will be reviewed in chapter 6.

We acknowledge here, as elsewhere, that a viability assessment can be a double-edged sword (S. Johnson, pers. comm.). If a PVA or other assessment concludes that no planning alternative can provide a reasonable probability of persistence for a target species, landowners and planners may be attempted to "write off" that species. This would usually be a mistake, because even small incremental gains in viability often can keep a population going (see chapter 4), and there are several examples of tiny populations, even smaller than 50 individuals, persisting for a long time (e.g., Walter 1990). In other cases a species may be declining to extinction for a variety of reasons unrelated to habitat conditions in the planning region, and even an ideal plan could do nothing to avert that trend. Planners must recognize that PVAs and other viability assessments are only tools and their predictions are often only crude estimates—useful, but necessary to weigh against other considerations.

Addressing Ongoing Threats

Does the plan contain specific mechanisms to reduce threats to each target species, its habitat, or the ecosystem generally? This question should be asked with regard to *in situ* protection, rather than relying on captive breeding to maintain population viability. In addition to the restoration and management concerns mentioned above, a plan should contain prescriptions for regulating human uses, controlling feral and exotic species, and, where possible, managing populations of potential predators, competitors, parasites, and mutualists. The specific mechanisms and management actions required will vary tremendously among species and sites, hence the desirability of detailed, site-specific studies to document factors affecting habitat quality and population viability. We cannot emphasize too strongly the need for fieldwork by competent biologists to determine the distributions and life history needs of species and the forces that threaten them, in order to create effective management plans. Because many threats cannot be foreseen, a plan should contain an adaptive management mechanism that will allow managers to recognize and address threats as they arise.

Development of Reserve Design

How adequate is the reserve design or other elements of the conservation plan in light of current theory in conservation biology, experience elsewhere, and factors specific to this species, habitat, or ecosystem? This question can be answered, first, by evaluating the proposed plan and its alternatives in light of the principles of conservation biology offered in chapter 4 and, second, by considering the species-specific and site-specific factors that make each case a special one. Because species- and site-specific information will never be as complete as desired, some reliance on general principles and experience elsewhere (e.g., with related or ecologically similar species) is inevitable and advantageous. A plan that fails to incorporate valid principles of conservation biology should be rejected. Plans that pay lip service to ecological principles, use the latest scientific jargon, and apply GIS and other modern technologies in impressive ways but nevertheless fail to incorporate these advances meaningfully in reserve design or management prescriptions, similarly should be rejected.

The Rigor of Science in Conservation Plans

The application of knowledge about the biological requirement of a species—or the management requirements of an ecosystem—should ultimately contribute to recovery. Often interim conservation plans, or planning agreements such as those in NCCP, are constructed in the absence of adequate knowledge about distributions of populations, population sizes, life histories, responses to natural or anthropogenic disturbances, and many other factors. We have already stressed that the conservation planning process should conform to established principles of conservation biology. However, detailed data on the species, natural communities, and sites involved are also needed.

Thus, a research agenda is a crucial component of most conservation plans (see chapter 3, box. 3.1). Frequently, the key questions for planning can be researched adequately in a period of two to three years, well within the time needed to develop most regional habitat-based conservation plans.

Research should concentrate in two areas: answering questions needed for planning and supplementing knowledge necessary for management. Research should be linked to monitoring in an adaptive management framework so that the effects of various management treatments can be assessed. For candidate species, research conducted in conjunction with a conservation plan can provide data useful in listing decisions. For example, a conservation plan that addresses the needs of a candidate species in a truly substantive way may obviate the need for listing the species in the future. Research will assist in addressing recovery goals for listed species by revealing more about life histories, habitat requirements, demography and genetics, and threats to the population. For communities and ecosystems, research is needed to determine interspecific relationships and provide information on disturbance regimes, succession, hydrology, and other ecological processes that affect the integrity of the system. Some questions for evaluating the rigor of science in conservation plans are considered below.

Integration of Science into the Planning Process
At what point does science enter into the planning process? Preferably, science is involved in all phases of conservation planning. The least useful stage is when scientists are called upon to judge completed plans or to second-guess final decisions that should have been informed by scientific information much earlier in the planning process. This does little more than create controversy and relegate science to the role of a competing interest in the negotiation instead of the foundation for conservation plans. We do not mean to imply that completed (but draft) plans should not be evaluated by scientists not involved in the plan. Rather, we insist that such plans then have an opportunity to be revised and, also, that independent scientists be involved as reviewers much earlier in the planning process. We suggested in chapter 3 that less analysis needs to be involved in cases that demonstrably pose less risk to species or ecosystems, such as HCPs for small projects involving very few individuals of a listed species on a site that has little potential for long-term viability under any scenario.

(As noted there we do not generally consider these small plans to be true habitat-based conservation plans, so we do not treat them thoroughly in this book. However, conservationists and planners in a given region should be wary about the potential for cumulative effects of many small projects.) Proper attention to site and population survey methodology and other data gathering, experimental design, monitoring, and ecological management is critical in all cases.

To assure that the highest standards are applied in all phases of conservation planning, input from scientists with appropriate expertise should be included at each step. At least some of these scientists should be independent from agencies, industry, or environmental groups. Ideally, such input should be as objective as possible. Although a science that is entirely objective and value-free is a mythology (Shrader-Frechette and McCoy 1993), science has well-established procedures for reducing bias, making assumptions explicit, and applying methodologies that can be duplicated by others (see chapter 4). Purely subjective input, regardless of the source, is often not constructive and can be counterproductive. The personal opinion of a scientist with regard to a particular conservation plan is not necessarily worth more than the personal opinion of any other citizen. On the other hand, subjective opinions of people—scientists and otherwise—who are personally familiar with a site or species should not be automatically discounted; at the least, these opinions provide hypotheses to be tested. Hypotheses and opinions, however, must be clearly separated from scientific evidence and informed inference.

Funding for Science

What is the level and continuity of funding provided for inventory, research, monitoring, and management? Although monitoring and management are often considered as features of plan implementation, rather than planning per se, we feel that a rigid separation of planning and implementation is artificial—the two should be part of one continuous and integrated process with multiple feedbacks (Noss and Cooperrider 1994). Science must be applied without interruption throughout the process, so that adaptive management can be tied to the results of research and monitoring. Some conservation plans are ostensibly science based, but contain little or no funding for site surveys, research (e.g., demographic or genetic studies of target species), ecological management, or monitoring. Without adequate, long-term funding for these activities a plan cannot be considered valid.

Research Agenda and Design

Are critical research questions and hypotheses identified clearly in the plan? Vague statements about research topics and information needs will not suffice to guide the acquisition of data in a rigorous way. Reserve design and management problems must be identified explicitly, or at a minimum their identification must be mandated, along with associated research questions and hypotheses. We suggest that research questions be placed in priority order and be followed by general suggestions and standards for methodology and estimates of time, staff, and funds required to conduct the research. These questions, suggestions, and standards should be part of the input to the planning process by scientific experts.

Map-based Analysis

Does the plan use maps and spatial analysis to assist reserve design and management decisions? Are all data layers for each alternative digitized and in a GIS at an appropriate spatial scale(s) for decision making? It is imperative today that conservation plans be map-based and accessible in GIS, so that plan alternatives can be displayed graphically and evaluated using quantitative information. The technology of remote sensing and GIS has advanced rapidly over recent years, and plans must take advantage of these advances (see chapter 6, box 6.2). Digital map data should be made available to other researchers upon request, so that they can conduct independent analyses to support implementation, critique plans, or offer alternatives.

The Contribution of a Plan to Community or Ecosystem Conservation

Conservation biologists increasingly recognize that planning based on the needs of single species or focused on small sites usually is not defensible because species and sites do not exist independently from other species and the broader landscape context. Equally as important, species-by-species planning fails to provide assurances to private landowners and other participants that they will not have to repeatedly come back to the bargaining table. Situations will still occur where a single-species or small-area plan is preferred or required, but we expect such cases to be increasingly rare. Even planning exercises stimulated by the presence of a single listed species will usually need to take into consideration candidate species that occur in the same area, other species that are locally rare and of biological interest, game species of value to the public, ecological factors that determine habitat quality, and many other concerns. Because the needs of no two species are identical, single-species plans for the same area can be in conflict if not closely coordinated. The extent to which a conservation plan takes into consideration multiple species and the ecosystem or landscape as a whole is an important criterion for assessing its potential for ultimate success. Some questions by which to evaluate the contribution of a plan or option to community or ecosystem conservation goals follow.

Attention to Species or Ecosystem
Does the plan explicitly consider multiple species and the broader ecosystem, or does it do so only incidentally? As noted earlier many "multispecies" plans are little more than a bundle of single-species plans lumped together. Such plans have the advantage of addressing several species at one time, but they may not be well coordinated or consider the interactions of these species with each other and with their physical environment. Community-level plans, such as the NCCP for coastal sage scrub in southern California, should properly address a suite of species and their collective habitat requirements, as well as ecological processes the ecosystem as a whole requires to be healthy. In practice, such plans may effectively focus mostly on a few target species whose needs together largely define the ecosystem (e.g., the California gnatcatcher, orange-throated whiptail [*Cnemidophorus hyperythrus*], and coastal cactus wren [*Campylorhynchus brunneicapillus sandiegoense*] in the case of California coastal sage scrub) or for which considerable data are available (e.g., the Florida scrub jay).

The target-species approach is not only inevitable in many cases, it is desirable in the sense that community- or ecosystem-level plans that fail to take into account the needs of the most sensitive species, those with the largest area requirements, or those that play critical roles in the ecosystem, will likely fail to maintain populations of these species (see box 5.1) and may place the community or ecosystem at risk. In many cases the spatial bounds of an ecosystem plan should be determined by the species with the largest area requirements (Noss 1996b). But if the needs of many other species or the integrity of the community are ignored because of lack of data or some other convenient reason, there is a danger of falling short of a community or ecosystem plan. Thus, natural community plans should be evaluated as to how thoroughly they address the broader community or ecosystem, at least through a comprehensive set of target species such as that offered in box 5.1. Single-community plans also have limitations. The ideal planning approach is probably one that focuses on the regional landscape as a whole—that is, the entire mosaic of natural and human-altered biological communities (Noss 1993, 1995, 1996b; Vance-Borland et al. 1995/96).

Umbrella Species

Does the plan develop an umbrella species rationale? Is the umbrella assumption well justified? How many other species would be covered by the proposed umbrella or set of umbrellas? How many would be left out? Have appropriate analyses been conducted to answer these questions (see Noss et al. 1996)? The argument that species with large area requirements serve as effective umbrella species, because areas required to meet their needs will encompass the areas needed by other species, is used frequently in conservation proposals. However, the umbrella species concept has been poorly tested. In surveying the literature Noss et al. (1996) could find no definitive, published studies documenting the level of protection afforded to other species by a conservation plan focused on any ostensible umbrella species. The President's Forest Plan (Option 9) for the Pacific Northwest was driven largely by two or three umbrellas—salmon and northern spotted owls, plus some attention to marbled murrelets *(Brachyramphus marmoratus)*—with the assumption that 380 + other species associated with late-successional forests in the region would be protected by these umbrellas (D. Wilcove, pers. comm.). The rigor of that assessment and the validity of the assumption were questioned by many scientists, and the impacts of the plan on the 380 + species are, with rare exceptions, not being monitored.

It seems reasonable that species with enormous area requirements should function as umbrellas; however, it is unlikely that any single umbrella species will shelter all other species, particularly in areas with high species richness and many endemics. In other cases a species closely associated with a certain natural community might serve as an effective umbrella for other species associated with that community that have similar ecological requirements. Even in such cases, however, a single species is unlikely to perform an umbrella function completely.

We expect that an assemblage of species generally will serve an umbrella function better than any single species (see box 5.1). For canyons in southern California, where endemism is high, 62.5 percent of the available area is needed to represent all bird, mammal, and plant species at least once (Ryti 1992). In this study plants (collectively) were considered the best umbrellas, because reserves representing all plants would capture 96 percent of the vertebrates (Ryti 1992). Ryti's study, however, did not consider population viability or area requirements for particular species. In north temperate regions with comparably low species richness and endemism, setting aside areas big enough to maintain viable populations of a guild of large carnivores, which collectively inhabit most of the natural communities of the region, might very well protect most other species. Considering just one species, the grizzly bear *(Ursus arctos horribilis)*, recovery zones proposed by Shaffer (1992) cover 34 percent of Idaho, in addition to large portions of adjacent states. These zones are not proposed to be exclusive of human activity; they are simply defined as the area necessary to allow recovery of grizzly populations. Two-thirds (67 percent) of the vegetation types and 65 percent of the native vertebrates in Idaho would have 10 percent or more of their statewide distributions included in the Shaffer plan; however, some species, such as reptiles, would be poorly represented (Noss et al. 1996) (fig. 5.1). Analyses at least this detailed should be undertaken before claiming umbrella species functions in a conservation plan.

Species Coverage
Are all listed, candidate, and other highly sensitive species considered adequately in the plan? This is not a simple question. Plans should be evaluated for how thoroughly they consider the needs of all species known to be or potentially vulnerable to human activities in the region of concern. This requirement is the root of assurances for the private sector, in that without consideration of all species likely to be vulnerable to human activity, a return to the planning table in the future is likely. In many instances the starting point will be simply a list of such species. Such a list should include all species recognized as endangered, threatened, or of special concern by state and federal agencies, as well as those listed by state natural heritage programs,

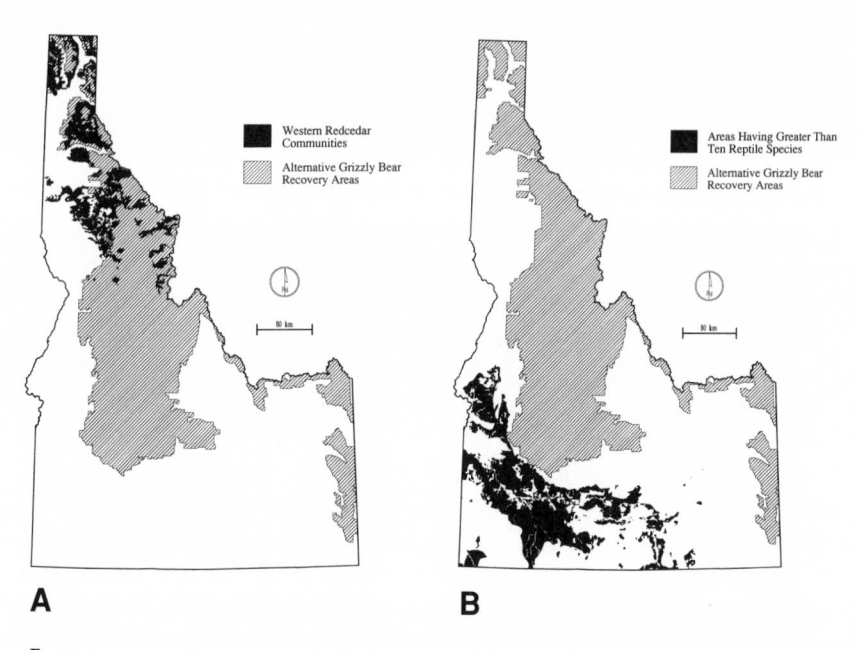

FIGURE 5.1. Alternative grizzly bear recovery zones proposed by Shaffer (1992) provide a good "umbrella" for many plant communities and associated species in Idaho, such as western red cedar forests (A), which are threatened by logging. But the proposed recovery zones provide a poor umbrella for arid-land habitats and species, including the centers of species richness for reptiles in the state (B). The lesson: no umbrella is complete. From Noss et al. (1996).

native plant societies, and other legitimate organizations. Even more useful might be a list constructed as suggested in box 5.1.

Concern has been expressed about the inclusion of species for regulatory coverage and assurances when their vulnerability and status is poorly known. We agree this is a valid issue. For instance, if inclusion on a list of "special species" prevents a species from being listed later under the ESA or, if it is listed, permits a level of incidental take that is not sustainable, the species may be placed at risk of extinction (M. Soulé, pers. comm.). On the other hand, it is important to identify species potentially threatened by development, and it would be unrealistic to expect a community- or ecosystem-level conservation plan to assess thoroughly the life history needs of all such species, much less include detailed assessments of their likely responses to plan alternatives. Also, plants are unlikely to be threatened by inclusion on a list of special species for conservation planning on private lands, even if this precludes their listing under the ESA, because plants are not protected from taking on private lands anyway.

As noted earlier the longleaf pine–wiregrass ecosystem of the southeastern United States has some 27 listed species and 99 candidates (before the FWS's elimination of the C-2 category) associated with it (Noss et al. 1995; Noss and Peters 1995). An ecosystem-level plan for this vegetation type, which comprises a broad assortment of plant communities, would, we hope, take care of most of these species without the need for detailed species-level recovery plans. Of course, in order to test this assumption some degree of species-level monitoring is needed; the more, the better, or otherwise some species could slip through the cracks. In all cases we suggest that, at least initially, truly questionable species be left off the regulatory coverage list with the explicit option of adding them later as information is gathered to substantiate their inclusion.

Ecological Processes
Has the plan considered critical ecological processes (e.g., nutrient cycling, erosion, soil building, weather and climatic influences, fire and other disturbance regimes, succession, pollination and other mutualisms, herbivory, predation)? Are there explicit provisions for allowing or enhancing the continued operation of processes essential to the ecosystem and mitigating disruptive processes related to human activities? Consideration of ecological processes is conspicuously absent from many conservation plans, especially at smaller scales, except for some attention to fire management for community types known to be sensitive to or dependent on fire. Yet, in probably most cases the nature of a biological community is determined more by physical conditions (climate, substrate, etc.) and processes (especially disturbance and hydrological regimes) than anything else. Ultimately, the biodiversity of a region depends on the continued operation of natural processes. Many of these processes require considerable space to function normally—that is, to operate within the natural range of variability that species have adapted to over evolutionary time. More attention to ecological processes in plans is urgently needed.

Spatial Scale of Analysis
Have the appropriate spatial scales for analysis, planning, and management been identified? Multispecies and ecosystem plans can be expected to require analysis at several spatial scales. Because many

endangered species are distributed as spatially subdivided populations (metapopulations, generally speaking), spatially explicit metapopulation models have become a popular kind of PVA for many species (see chapter 6). These models are complicated to conduct and are quite data-intensive. The technology is rapidly developing in this area, however, and some kind of spatially explicit models for target species should guide both planning and adaptive management. The scale of analysis and planning area ideally should correspond to the demographic and genetic structure of the most wide-ranging species (Merriam 1991) or to the scale set by natural processes, such as disturbance and hydrological regimes. Often a conservation plan, even for a single species if it is one with a complex demography, requires analysis at several scales. Useful map scales might be 1:100,000 for regional analyses, 1:24,000 for landscape (subregional) analyses, and 1:4,800 or even finer for site-level analyses. Thus, the plan as a whole should be analytically hierarchical. Funding must be adequate to support this level of research.

Temporal Scale of Analysis
Have appropriate temporal scales for analysis, planning, and management been identified? Some ecological processes (e.g., the life cycles of annual plants or mayflies) cycle over very short periods of time, whereas others (e.g., stand-replacing fires, droughts, disease epidemics) may cycle on scales of decades or centuries. Populations respond in complex ways to cyclic or episodic events. For example, shrubland bird populations vary tremendously in size over time, corresponding to variation in precipitation and loosely tracking rainfall–drought cycles, but often with time lags associated with site tenacity and other poorly understood factors (Wiens 1986). Florida scrub jay populations are dramatically affected by avian virus epidemics that may occur only once every 25 to 40 years (J. Fitzpatrick, unpublished data). Planning and management will have to account for these periodic, stochastic, and cyclic events or risk losing species.

Ecosystem management is largely untested. As noted earlier, ecosystems and their responses to management are extremely complex. But this is no excuse for failing to use the best available ecological theory, models, and empirical information to construct plans and adaptive management strategies. The public should be able to expect a reasonably high degree of scientific sophistication in multispecies, community, and ecosystem conservation plans. Consideration of disturbance regimes, for instance, and responses of different species to these regimes is quite feasible for most ecosystem types in the United States. Though we lack detailed information on species-specific responses in many cases, enough is known to develop reasonable management guidelines for allowing natural disturbances, or where this is impossible, for simulating their effects through prescribed fire, silvicultural manipulations, or other techniques.

The Adaptive Management Capability of a Plan

Adaptive management is perhaps the most important issue for the implementation phase of conservation planning. Because habitat conservation plans will always be experiments with uncertain outcomes, some elasticity is required in implementation so that managers have the advantage of learning from experience and modifying their practices accordingly. This flexibility need not reduce the level of regulatory assurances received by the private sector, and it is a fundamental element in the success of the plans. One of our principles (see chapter 4) is that nature is full of surprises. The implementation framework for a plan should be flexible enough to respond to unforeseen circumstances. If not foreseen, new circumstances must at least be seen, which requires a commitment to ecological monitoring. This monitoring, using quantifiable indicators and placed in a hypothesis-testing framework with a valid experimental design, is essential for measuring progress toward conservation goals and comparing the long-term effects of management practices (Noss 1990; Murphy and Noon 1992; see chapter 6). In order to learn from experiments, management must be adaptive (Holling 1978; Walters 1986; Walters and Holling 1990).

A conservation plan requires a long-term obligation to ecological monitoring and to adjusting plans on the basis of new information. For example, the monitoring plan for the Coachella Valley fringe-toed lizard has uncovered, over the past decade, a number of important factors affecting both lizard populations and the physical processes of the system that are crucial to implementing the plan (Barrows 1996). The lizard population has been monitored within three preserves since 1986; the results of these surveys indicate that lizard populations fluctuate with the availability of loose sand, insects, and other resources. Monitoring only lizards, however, would not permit adaptive management—it would be "an academic exercise, with no options for remedial protection efforts" (Barrows 1996:890). Through the use of a valid ecological model, monitoring in this case was focused on sand sources and wind corridors critical to maintaining the sand-dune habitat of the lizard. The HCP for the lizard is being revised to reflect this new knowledge and protect additional habitat.

Unfortunately, relatively few conservation plans of any kind have well-developed and statistically valid monitoring programs (Noss and Cooperrider 1994). In fewer cases still has monitoring been implemented adequately and carried out over a period of years so that trends can be noticed and responded to. In many cases recovery plans, HCPs, and other conservation activities proposed under the ESA either lack monitoring programs altogether or have extremely vague requirements for how plans should be modified on the basis of data derived from monitoring. Under these circumstances learning from experience in any rigorous way is impossible.

Private landowners, developers, and extractive industries participate in habitat conservation planning because of the promise of regulatory certainty and assurances regarding their future activities and obligations. As we have discussed, the "no surprises" policy of the U.S. Department of Interior is an attempt to respond to this need. The intent of the policy is to encourage landowner cooperation in conservation planning by offering assurances that they will not be continually asked to invest more in the conservation plan at some later date. There has been tremendous controversy over the issue, however, because assurances in plan agreements may run counter to adaptive management if they are regarded carelessly.

A no surprises policy should under no circumstances be interpreted to mean that we can predict all the consequences of implementing a particular conservation plan. New information on the status of species and communities and their requirements for long-term viability, as obtained through monitoring and research, can be expected to lead in many cases to changes in plans, including in some cases an increase in the amount of land protected. If opportunities for modifying and improving plans on the basis of new information are precluded, failures in attaining biological goals are likely (Barrows 1996). Of course, some opportunities for modifying plans will be precluded by developments occurring after approval of the initial plan—if a highway or subdivision is built, it will be difficult to remove. Such permanence is legitimately of great concern to the environmental community. On the other hand, underpasses and fences for wildlife movement can be built under highways, habitat might be protected or restored at sites other than where the subdivision now occurs, habitat management might be intensified, and so on. It is not unreasonable to promise private participants that most if not all of the financial burden of additional requirements should be borne by society as a whole. But we should not confuse financial assurances to participants with the flexibility needed to adapt conservation strategies to the uncertain behavior of ecosystems. The following questions address the adaptive management potential of a conservation plan.

Monitoring Program
Does the plan or alternative include a scientifically and statistically valid monitoring program? Monitoring is frequently ignored in implementing conservation plans, NEPA projects, and land management activities generally (see above). Hence, opportunities for learning about the effects (positive or negative) of various activities and modifying the plan are extremely limited. Monitoring need not be highly complex and expensive—if too expensive, it often would be better to put the money into additional land acquisitions (S. Johnson, pers. comm.). Recommendations for developing monitoring programs as an essential component of adaptive management will be provided in chapter 6 (see also Noss and Cooperrider 1994).

Testing Management Practices
This category is closely tied to the one above. One should not monitor for monitoring's sake, but to test hypotheses and inform management. Does the plan allow for testing of hypotheses regarding effects of management practices on populations and other conservation elements of concern? Does it allow for testing of alternative management treatments? Rigorous testing of hypotheses requires a valid experimental design, including true replicates and control areas. True replication is not possible in many cases of applied science, especially in landscape-scale experiments. Although lack of replication limits opportunities for statistical inference and extrapolation, unreplicated experiments often yield results obvious and striking enough that much can be learned from them (Hurlbert 1984). "Natural experiments"—for example, differing growth rates of populations in natural areas of different sizes, degrees of isolation, or landscape contexts—should be taken advantage of whenever possible (Diamond 1986).

Because of our general ignorance about how to manage natural communities, alternative management treatments should be tested, with care taken to avoid or rapidly terminate treatments that appear to have lasting, negative effects on imperiled species or ecosystems. Short-term negative effects can be expected and need not pose a problem; for example, a prescribed burn might kill many individuals of a target species, while enhancing habitat quality and increasing carrying capacity for the species over the long term. A plan that fails to test alternative management regimes in terms of their effects on elements of biodiversity offers few opportunities for adaptive management.

Timely Analysis

Does the plan include a mechanism for regular and timely analysis and review of monitoring data? Many agencies have been accused of "blind data gathering," collecting masses of information but then never making an effort to analyze the data and interpret them. To avoid this problem plans should include a specific timetable for analyzing and interpreting monitoring data in order to inform management decisions.

Responsiveness to Information

Is the plan designed to be responsive to information derived from monitoring? To what extent can the plan be modified to take into account new information? As noted above, a conservation plan "set in stone" and designed to avoid future surprises is an inflexible plan that potentially places species and ecosystems at great risk. Nature is dynamic and unpredictable. Surprises are bound to occur; it is only a question of whether we notice them or not. The sooner we notice them and take corrective action, the lower the risk to biodiversity. Thus, plans should be evaluated as to how open they are to modification based on new information. Plans must be dynamic and explicitly built on a foundation of adaptability and revision. We recommend that review and reconsideration of management practices should be contractually obliged in plan implementation agreements.

Appendixes 5A through 5C illustrate some examples of conservation plans, pointing out their strengths as well as their weaknesses. These case studies are by no means exemplary of the range of plans completed or under preparation, but they are meant to illustrate some of the challenges alluded to in this chapter and throughout this book.

APPENDIX 5A

Orange County Central/Coastal NCCP Case Study

Southern California has the most extensive experience with habitat conservation planning of any region and is responsible for most of the innovation and evolution of the HCP mechanism. The prototype conservation plan was developed in 1982 for remaining habitats on San Bruno Mountain on the San Francisco peninsula. That same year HCPs were included as amendments to Section 10(a) of the federal ESA (Bean et al. 1991). Partly due to the work of a core of individuals in government and the private sector with experience in HCPs, conservation planning has grown steadily in California over the past 15 years, to the point that large expanses of the region are under some type of conservation planning process.

With extensive use of HCPs has come a realization of their limitations in achieving the goals both of conservationists and the regulated community (O'Connell and Johnson 1997). Perhaps the greatest overall shortcoming is that HCPs are a mechanism to resolve conflicts created by the prohibition on incidental taking by modification of the habitat of listed species on private lands (a prohibition recently upheld by the Supreme Court). Those prohibitions only extend to listed animal species and occupied habitat, quite a bit less than what is necessary to conserve biodiversity, as we have described throughout this book. (HCPs are not developed directly for plants because incidental take—indeed, any take—of plants on private lands is not prohibited.) The practical standard that has been applied to approval of HCPs is that the plan "not appreciably reduce the likelihood of survival and recovery of the species in the wild" (i.e., the jeopardy standard). This means that a plan that brings a species closer to extinction may still be approved as long as it doesn't push the species over the brink.

There are other limitations to HCPs—and the ESA—from the conservation perspective, including little opportunity for public participation in the development of private landowner conservation plans and a historic lack of credible, independent scientific input into many HCPs (O'Connell and Johnson 1997). But the limitations are not only to conservation. The regulated community has been frustrated by the lack of long-term certainty that planning for only listed species brings, as well as their perception that a permitting process focused on private landholdings places an unfair share of the financial burden on them. Indeed, although the FWS has recently begun to encourage larger-scale planning processes under Section 10, most of the nearly 400 plans approved or in process at this writing have been for single ownerships. In fairness, since HCPs are designed to respond to the potential incidental take conflicts set up under Section 9, most of these problems stem from the ESA itself.

As we discussed in chapter 2, the State of California in 1991 unveiled a program designed to address the limitations of HCPs both for conservation and for private property owners. The Natural Community Conservation Planning Act created a legal mechanism to enable landscape-scale planning and conservation. The pilot project for NCCP was the coastal sage scrub complex of natural communities in coastal southern California (fig. 5A.1). In exchange for providing far more habitat conservation than the federal ESA mandates for the numerous rare and endemic species, both listed and unlisted, that occupy the region, this voluntary program promised landowners far greater long-term regulatory relief than has been possible under the federal ESA (O'Connell and Johnson 1997). This relief is what troubles environmentalists, who fear that the ESA is being undermined. For practical purposes, the 6,000-square-mile planning region was divided into nine subregions, and plans were and are being developed by local governments for each region. The Central/Coastal Orange County subregion was the first plan attempted and is the subject of this brief case example.

Orange County, California, can be characterized as a place of extremes. Extreme development pressure has eliminated most of the native habitat and threatens most of what remains. Its standard of living, among the highest in the country, has led to extraordinary property values—some undeveloped quarter-acre lots with ocean views sell for millions of dollars. It is also a place of extreme politics, both liberal and conservative, reflecting a highly polarized public.

The Orange Central/Coastal NCCP is a good example of one of our principles from chapter 4 that the less cumulative impact a project involves, the less scientific scrutiny is necessary in reserve design. The Central/Coastal NCCP subregion lies in the heart of one of the most urbanized areas of the United States. Of the 208,700 acres within the planning boundary, more than 135,000 were either already urban, in intensive agriculture, severely disturbed nonnative grassland, or in-

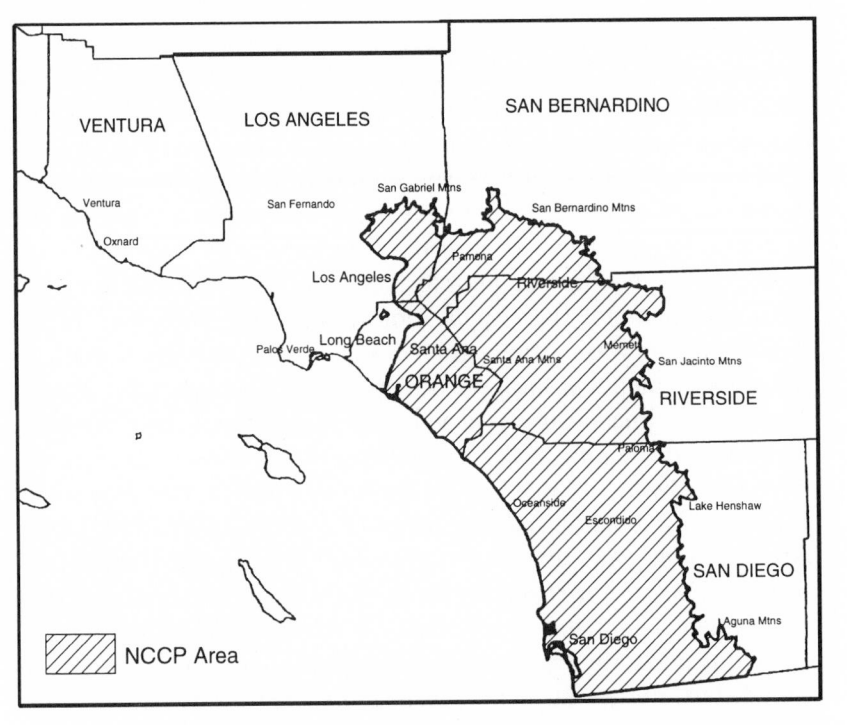

FIGURE 5.A.1. The Natural Communities Conservation Planning region for the coastal sage scrub ecosystem in southern California. The area includes most of the landscape on the coastal side of the mountains below 3,500 feet, within the area of southern California inhabited by the California gnatcatcher and many other endemic species.

side the Cleveland National Forest. Of the 73,700 acres of natural and seminatural habitat that remained, roughly 15,000 acres were highly fragmented and significantly degraded from prior impacts. Through a combination of preserve designation for most of the remaining habitat and continuing strict ESA protection on other parcels, the NCCP protected or will lead to the protection of approximately 85 percent, or 50,000 out of the 59,000 biologically valuable acres (County of Orange EMA 1996). This solution is unsatisfying to many people, for different reasons, because it begs two opposing questions: In a landscape so developed, isn't every undeveloped acre important and worthy of protection? Or, conversely, in a landscape so developed, does it really matter whether more development occurs? Perhaps, in a world of limited funding, conservation efforts should be focused on more pristine landscapes, where chances of success are higher. On the other hand, the coastal sage scrub community is among the biologically richest and most threatened in the country, and thus scores high in terms of conservation priority.

The overall NCCP program is founded in a set of conservation guidelines that were created by a panel of independent scientists, the Scientific Review Panel (California Department of Fish and Game 1993). These guidelines amount to a series of tenets for reserve design and instructions for their application to planning. The intent of the NCCP program was for these guidelines to inform development of the biological elements of local plans. Although a set of NCCP planning process guidelines also was developed by the state, their direction amounted to enabling each local government to pursue its planning the way it saw fit. Orange County did so by hiring a consulting team of planners and biologists to develop the reserve design alternatives. This approach generated controversy, both locally and from distant observers, because it relied on relatively little input from outside or independent scientists. Superficially, at least, the plan appeared to many much like the traditional HCP processes, on which NCCP was intended to improve by making plans more scientifically rigorous and credible. The analysis for the Orange Central/Coastal plan, however, was fairly straightforward and followed the tenets of reserve design from the conservation guidelines reasonably well.

The reserve design process in Central/Coastal Orange County was forced to concentrate on creating and retaining as much landscape connectivity as possible. Most remaining habitat was aggregated in two large, disconnected blocks, one on the coast and one inland. These cores are separated by, among other things, two ten-lane freeways and several miles of suburban development and intensive agriculture. Because of the history of local land use and the urbanized nature of the landscape, most of the remaining habitat was or had been considered for open space designation previously. A large portion of the coastal core area was public land, and approximately 21,000 acres of the remainder belonged to a single, large landowner. Because of these considerations, the reserve design process focused primarily on linkages between habitat blocks.

To evaluate linkages for the reserve system, the consulting team examined each core area and potential landscape connector, adding or subtracting acreages as they deemed appropriate. Some landscape links were narrowed and others were expanded based on the judgment of the consulting biologists. In the end roughly 3,000 acres had been added to the existing, proposed set-asides for connectivity purposes. The total strict NCCP reserve was 38,738 acres.

Despite the attempts of planners to adhere closely to the reserve design guidelines, some were violated to varying degrees. For example, two basic tenets of the NCCP program are that roadless or otherwise inaccessible areas provide better protection to target species than areas with roads, and that habitat fragmentation should be avoided. In the coastal area, a controversial major transportation artery was allowed to bisect the large habitat block, dividing the reserve into two pieces of roughly 7,000 to 10,000 acres each. However, the regrettable decision to build the highway was not under the control of NCCP planners, but was allowed by a Section 7 consultation in advance of the NCCP (S. Johnson, pers. comm.). The consultation allowed highway planners to mitigate impacts of the highway through the NCCP. As additional mitigation, which many biologists believe is inadequate, three wildlife underpasses were constructed along the road.

The Orange County Central/Coastal NCCP is also an important case study of the need for long-term, adaptive management principles and requirements in conservation plans. Given the condition of the reserve lands, their urban character, and the likelihood of continuing external threats, the scientific elements of long-term adaptive management are more crucial in this conservation plan than most. The approved NCCP, however, did not contain a detailed adaptive management program. Instead, it mandated the creation of such a program in the future with a rigid timeline. In particular it identified three required elements of the program: (1) restoration, enhancement, and exotics control; (2) fire management; and (3) public access management. In addition, a grazing management plan was created to convert cattle grazing from a year-round activity to a seasonal impact that could be controlled. The NCCP stated that the adaptive management plan would be fully operational one year following approval of the permit, but provided little explicit guidance. At this writing the plan is being drafted and will be subject to approval by the FWS.

The land management program for this NCCP has some positive qualities. First, to comply with the terms of the implementation agreement, all lands within the reserve system, including existing public land, must be managed in a coordinated way with conservation as the highest priority. Much of this land historically suffered from lack of management. Intensive recreation such as mountain biking and equestrian use affected some areas, though most of the lands were not open to public use. The NCCP requires that all uses and impacts be brought into line with the biological goals of the plan. Second and equally important, the NCCP created a permanent endowment of $10.6 million for management costs, financed by the participating landowners and public agencies. These funds are to be used for the biological management of the reserve system and are in addition to the other management costs, such as fencing, security, and debris removal, that are also mandated by the plan and will be carried out by the owners of the land.

In public hearings the Orange Central/Coastal NCCP was controversial within the environmental community. Some local and national groups supported it and others adamantly opposed it. The landowners involved were in favor of the plan, which made some environmentalists dislike it even more. The NCCP developed a target species rationale, but sought little detailed biological evaluation of other species. The lack of independent scientific input beyond the general conservation guidelines was criticized by some opponents (this is largely an issue of distrust of agencies) and there were many calls for "peer review." Yet, the acreage of natural habitat to be lost from this already highly urbanized and fragmented landscape will be relatively small under the NCCP. The plan developed a clear and detailed biological and implementation monitoring program, but left the intricate adaptive management planning to the implementation phase with a large endowment. Although it is difficult to evaluate whether the plan will contribute substantially to recovery of the 37 species it covers, especially in the early years of implementation, the required management goal is a net gain in habitat value for the species over the long term. It remains to be seen how well the adaptive management program tracks the principles developed in this book.

APPENDIX 5B

Brevard County, Florida, HCP Case Study

As we have described, regional habitat-based conservation planning is a complex process influenced by countless biological, social, economic, and political conditions. When all these factors are reconciled, plans have a good chance of success. If, however, any major element is not resolved it may be difficult for participating interests to achieve their goals. The Brevard County, Florida, HCP is a good case study of a plan strongly founded in science (tracking closely our criteria in this book) and containing workable economic provisions, which fell victim to the extreme antienvironmental sentiment among many conservative elected officials during the 104th Congress. Not only does the HCP illustrate an exemplary scientific process, it demonstrates beyond doubt that proper application of science is not the only requirement for a viable conservation plan.

Surprisingly, given the rapid human population growth and the large number of listed species, habitat conservation planning has not been used extensively in Florida. An HCP was created for the tropical natural communities of North Key Largo in the mid-1980s, but that plan was never submitted for a permit. Instead, the Conservation and Recreation Lands (CARL) acquisition program of the state of Florida purchased the proposed reserves and development sites, obviating the need for an incidental take permit and HCP. A couple of tiny Section 10(a) permits were issued in the Florida Keys for single landowners in the late 1980s. Until 1992 this was Florida's only experience with habitat conservation planning under the Endangered Species Act.

Brevard County lies roughly 200 miles north of Key Largo along the central east coast of Florida and includes the human communities of Merritt Island, Melbourne, and Titusville, as well as the Merritt Island National Wildlife Refuge, the Indian River Lagoon estuary, and NASA facilities at Cape Kennedy (see fig. 5.B.1). It is a rapidly urbanizing jurisdiction that still harbors many of the rural attitudes about government regulation and private property rights that are at the core of the antienvironmental movement.

Brevard began its habitat conservation planning process in 1992, largely in response to the federal listing of the Florida scrub jay *(Aphelocoma coerulescens)* as threatened and subsequent movement by the FWS toward holding local officials liable for approving projects that destroyed habitat occupied by jays. After studying a number of potential responses to the listing, the local government decided that given the looming threat of a land-use crisis, the question of their own personal liability, and the knowledge that many more species in the scrub natural community could become listed in the future, a county-wide conservation and development plan for scrub was the surest way to resolve the conflict and regain regulatory certainty.

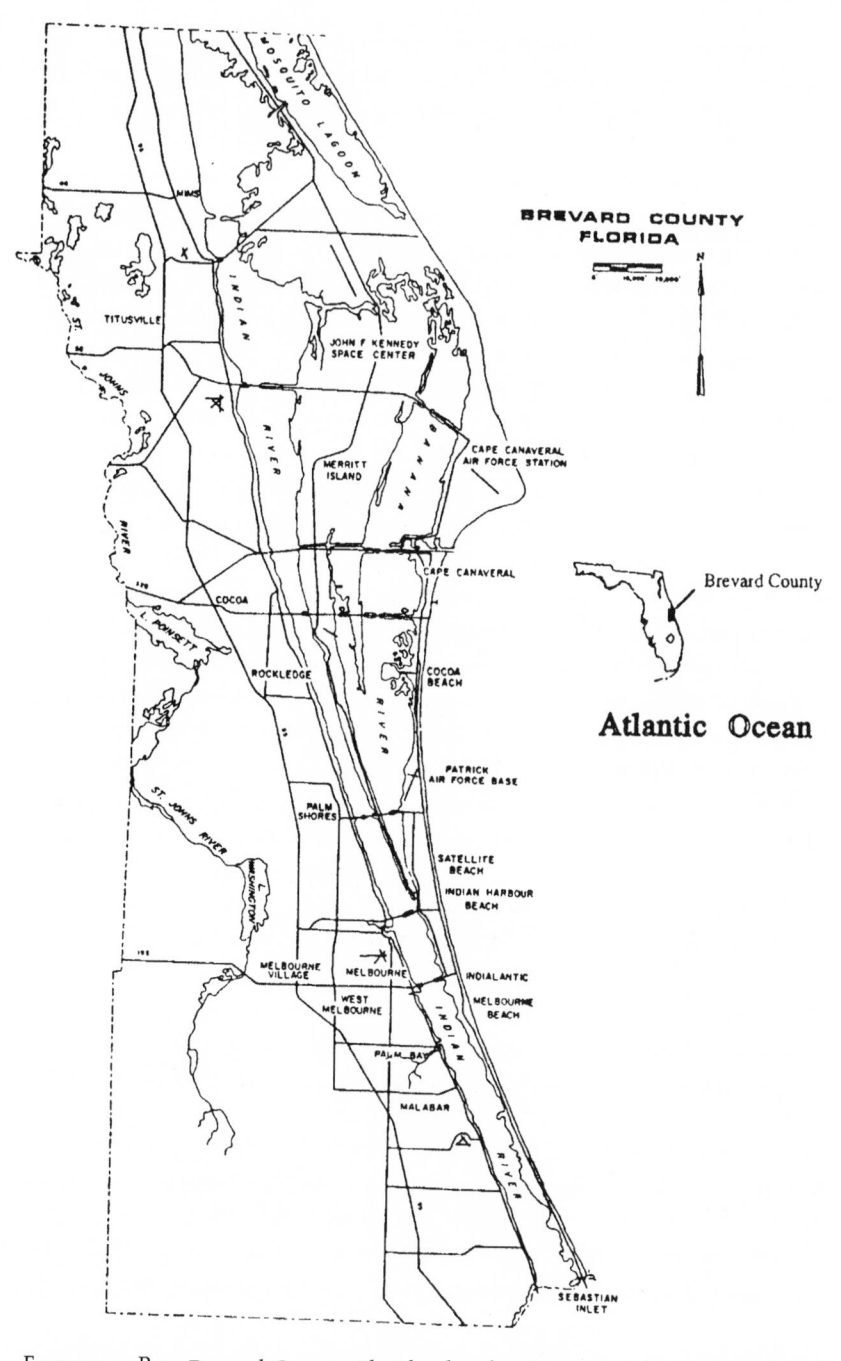

FIGURE 5.B.1. Brevard County, Florida, the planning region for a science-based
Habitat Conservation Plan that failed as a result of political opposition.

Building on the experiences of previous regional HCPs elsewhere in the country, the Brevard County process broke ground in many areas, including citizen participation, economic analysis, and the structure of the planning process. These elements provide interesting case studies in their own right and should be part of any general examination of habitat conservation planning. However, for the purposes of this example we focus our discussion on the scientific techniques and analyses used to develop the alternatives for the conservation plan.

Florida scrub is a complex of imperiled natural communities characterized by sandy, nutrient-poor soils, periodic drought, high seasonal rainfall, and frequent lightning-caused fires (Austin 1976; Myers 1985; Myers and White 1987). Scrub is found largely on ancient dune ridges left from oceanic transgressions (Laessle 1958, 1968); many scrub sites were islands during Pleistocene interglacial periods with high sea levels. Scrub was once one of the more prevalent habitats along the central ridge and coasts of the state. However, scrub is also one of the most prized habitat types for development due to its generally high-dry character and permeable soils. Estimates of loss of scrub over the past century run as high as 85 percent (Fitzpatrick et al. 1997). In Brevard County only 30 percent of historic scrub patches remain, and those are dramatically fragmented (H. Swain et al., unpublished). Further, suppression of the fire cycle has resulted in overgrowth of most remaining scrub and generally poor habitat quality (Givens et al. 1984). Florida scrub jays, for example, depend on midsuccessional scrub dominated by oaks and abandon sites when a canopy of sand pines forms after several decades of fire suppression.

The extreme environmental conditions and natural isolation of scrub patches has resulted in remarkable speciation, especially among plants (Christman and Judd 1990). Dozens of narrowly endemic species are found among the ridges. In Brevard County these include a newly described mint species *Dicerandra thinicola*, the large-flowered rosemary *(Conradina grandiflora)*, and Carter's warea *(Warea carteri)*, among others (H. Swain et al., unpublished). In addition to the scrub jay, rare fauna include the federally threatened Eastern indigo snake *(Drymarchon corais couperi)*, the Florida mouse *(Podomys floridanus)*, the gopher tortoise *(Gopherus polyphemus)*, and several cryptic reptiles and amphibians.

Brevard County has a history of action on local habitat issues that greatly helped inform the scientific process of conservation planning.

In 1990 local citizens passed a bond measure allowing up to $55 million for acquisition of critical habitats. The bond proceeds were administered by a group of citizens and scientists with technical expertise about endangered habitats. As part of their examination of properties to buy, and in advance of and independent from the HCP process, the endangered lands group conducted a preliminary study of scrub habitat in the county that identified key core areas for preservation. Acquisition was proceeding on several parcels before the HCP was initiated.

When the county began planning, it created two groups. The first was a steering committee of citizens representing stakeholding interests. Whereas this group included environmentalists, it did not contain any scientists. Previous scrub conservation efforts under the endangered lands program had revealed that designing a viable scrub conservation reserve was a highly complicated analytical task. So, the second panel created was a group of independent scientists charged with the task of developing reserve alternatives and advising the citizen group on matters of conservation science. The scientific panel included species experts and also ecologists with local expertise. The planning structure was extremely effective, for it allowed substantial scientific input into the decision-making process while maintaining the objectivity and independence of the scientists. The scientists themselves worked hard to maintain and protect this neutrality.

In addition to the advance work of the county endangered lands program, the scientists and citizens developing the Brevard County HCP had other advantages rarely found in endangered species conservation efforts. The first was that the Florida scrub jay is perhaps the best-studied bird in North America. More than 25 years of life history data, demographic measurements, and population modeling for the species were readily available (Woolfenden 1974; Woolfenden and Fitzpatrick 1984; Cox 1987; Fitzpatrick et al. 1991). Another advantage was that through long-term study, the scrub jay was known to be a good indicator or "umbrella" for the scrub natural community (Cox et al. 1987; Fitzpatrick et al. 1991; Fitzpatrick et al. 1997). Because of its broad area requirements and dependence on scrub, a conservation plan that worked for the jay would be expected to work well for most other species in the natural community. Another important advantage was the existence of a draft set of rangewide guidelines for scrub jay conservation commissioned by the state office of the FWS specifically to aid HCPs (Fitzpatrick et al. 1997). This relatively unprecedented and farsighted action on the part of the Service gave planners a clear picture of a biologically acceptable conservation plan. Usually, the FWS provides very little advance guidance for conservation of listed species through HCPs. Recovery plans can be helpful in some cases (see chapter 2), but the superficial recovery plan for the Florida Scrub Jay contributes little to the welfare of the species.

Despite the relative luxuries described above, the Brevard science advisors were immediately faced with a tremendous lack of locally specific data that would enable defensible reserve design. One scrub jay survey had been conducted by the county environmental resources division, but it was limited in scope and rapidly growing out of date (Snodgrass et al. 1991). Using a federal grant, the advisors hired a carefully chosen consulting team to gather four important and massive data sets. The data collected were:

1. Mapping of all distinct parcels of scrub habitat larger than two acres in size

2. A thorough map-based scrub jay population survey for the entire planning area including rough territory boundaries

3. Detailed field assessment of habitat quality variables for each scrub parcel potentially valuable for reserve design (a total of 1,200 separate patches)

4. Mapping and classification of nonscrub land-uses between patches of scrub habitat for the entire planning area.

These four data bases were developed as overlays in a GIS, enabling both digital analysis of reserve design alternatives and evaluation "by hand" using print maps. While data were being collected, the science advisor group developed a set of guiding principles for the reserve design process and tested a spatially explicit population model that would allow evaluation of reserve alternatives. The principles were developed using all information known about scrub and scrub jays, and they included both generally accepted tenets of reserve design (see chapter 4) and some concepts specific to scrub and scrub jays. The reserve design principles were (Swain, unpublished):

1. Bigger reserves are better than smaller reserves.

2. Reserves well distributed in the landscape preserve species better.

3. Connected or adjacent reserves are preferable to isolated reserves.

4. Shape of reserves is important, and incompatible adjacent land uses should be minimized.

5. Different intervening land-use types have varying effects on dispersal distance.

6. No local subpopulation of scrub jays currently containing more than 100 pairs should fall below that threshold as a result of reserve design.

7. No local subpopulation of scrub jays containing between 10 and 100 pairs should fall below two-thirds of its present size.

Once the GIS overlays were complete, the science advisors used these reserve design principles to analyze and interpret the data. By comparing parcel-specific data to the design parameters, each habitat patch was assigned to one of three categories: *biologically essential* (the core of the reserve system); *alternative* (allowing flexibility in alternative designs); and *released* (so fragmented or unrestorable that it would be impossible to protect as viable habitat). Using the spatial data, four alternative reserve designs were developed with the core reserve and combinations of parcels from the alternative category. The overall alternatives were tested through a preliminary spatial metapopulation model. Each alternative emphasized a different biological objective. The first (and largest) maximized overall viability of the ecosystem. The second minimized restoration cost by choosing sites based on ease of management. The third emphasized existing habitat quality of protected patches, and the fourth maximized connectivity between reserves. Each of the four designs met or exceeded the goals identified in the FWS's draft HCP guidelines for scrub jays.

The entire scientific process consumed 18 months and roughly $200,000. The four reserve alternatives were publicly presented to the citizen steering committee in early 1994 and generated little controversy. The controversy instead came over the policies and procedures identified by the steering committee to assemble the reserves. Opposition was firmly rooted in an activist private-property rights and antigovernmental movement that swept through the county following the election of the 104th Congress in 1992 and an antiregulatory majority on the newly elected county commission. In 1993 these local officials began dismantling their own environmental regulations, including development, landscaping, and wetlands protection ordinances.

Each reserve alternative was estimated to cost roughly $90 million with an annual management cost of $1 million. During the planning process, very little scrub habitat was developed, and all through expensive individual 10(a) incidental take permits. The steering committee commissioned an economic study that demonstrated a 75 percent savings per acre for the regional plan over individual 10(a) permits (Fishkind et al., unpublished). Despite this economic evidence a majority of local officials, expecting imminent emancipation from the ESA by Congress, were emboldened to postpone consideration of the HCP or any other alternative besides elimination of the federal regulatory burden. At this writing the Brevard Scrub HCP remains a thorough, scientifically well-developed approach to regional conservation and development that occupies shelf space in the county natural resources division. The conservation outcome of this case study is not all regrettable, however. The analysis and reserve design created by the science advisors and endorsed by the steering committee formed the foundation of a 1993 proposal to the Florida CARL program. In 1996 the "Coastal Scrub Ecosystem Initiative" ranked fifth on the state's top 100 list of acquisition projects for matching funds. Much of the scrub of Brevard County may become protected after all, though not nearly as efficiently as it might have been.

APPENDIX 5C

Coachella Valley, California, MSHCP Case Study

The Coachella Valley of southern California lies within Riverside County and contains Palm Springs and other famous resort cities with high development pressures. As discussed elsewhere in this book, the Coachella Valley was the site of one of the early HCPs, a single-species plan for the Coachella Valley fringe-toed lizard that was habitat-based and has shown a commendable degree of adaptive management capability through its application of an ecological model (Barrows 1996). Nevertheless, the HCP for the lizard, even with the designation of additional preserves since the original plan, protects little available habitat and will not assure viability for the dozens of other rare species of the Coachella Valley, many of which are found nowhere else. Hence, in late 1993 a scoping study was initiated to investigate the potential for a multiple-species HCP (MSHCP) for the valley. This plan is now in an early stage of development.

The scoping process in 1993 identified the need for and feasibility of a multiple-species plan. The first steps to organize the planning effort were taken in 1994, and in 1995 biological surveys began on public lands and on private lands for which permission had been obtained by landowners. As of 1996 participants in the MSHCP include the county government, nine city governments, the Coachella Valley Association of Governments, the California Department of Fish and Game, FWS, Bureau of Land Management, U.S. Forest Service, and National Park Service. In late 1996 a workshop was held to discuss progress on the plan and to peer review it; participants included many local experts and three prominent conservation biologists from outside the area as reviewers.

The proposed Coachella Valley MSHCP will consider a large number of species (19 main target species, at this writing, and a much longer list of Species of Concern) and many natural communities (23 "sensitive natural communities"); will give explicit attention to ecological processes (e.g., wind corridors and blow-sand sources); and includes both private and public lands in the Coachella Valley, the surrounding mountains, and some surrounding watersheds and adjacent desert lands. Thus, the proposed scope of the MSHCP is possibly broader than any habitat-based conservation plan to date, including NCCPs or even such ambitious efforts as the President's Forest Plan in the Pacific Northwest. Efforts are being made to get the MSHCP recognized as equivalent to an NCCP, thus qualifying the plan for increased federal funding. The process suffers from lack of sufficient funds, so fund-raising is currently the highest priority (see below).

The purpose of the Coachella Valley MSHCP is stated as follows (Coachella Valley Mountains Conservancy, unpublished):

> to conserve adequate habitat in an unfragmented manner to provide for the protection and security of long-term viable populations of the Species of Concern . . . to conserve biodiversity through protecting adequate areas of the natural communities occurring in the plan area to maintain functioning ecosystems and protect the species that inhabit them . . . [and] to proactively address requirements of the state and federal endangered species acts to avoid disruption of economic activities in the CV MSHCP area.

Four general approaches are proposed to achieve these goals: (1) protecting sufficient habitat for Species of Concern not currently listed so as to obviate the need to list them under the federal or state ESAs; (2) for Species of Concern currently listed under the ESA, use the MSHCP as the basis for securing incidental take permits within the area covered by the plan; (3) for Species of Concern not currently listed, address their conservation needs as if they were listed, so that if they are subsequently listed, no further mitigation requirements will be imposed for issuance of incidental take permits; and (4) utilize a habitat-based approach to conserve overall biodiversity in the planning area—specifically, by mapping vegetation, conducting a gap analysis, and attempting to assure that all natural communities are adequately represented in lands managed more or less for natural values.

Planners for the MSHCP have developed a list of initial reserve "design" criteria for selection of preserves, accompanied by a checklist of biological and physical processes to consider in making such decisions. The first criterion is whether a site has high biological value; the second set of criteria (for sites meeting the first criterion) is whether biological and physical processes can be sustained on the site; the third set (for sites meeting the second criterion) is that the level of threat is not unacceptably high or that threats can be successfully managed; and the fourth set (for sites meeting the third criterion) deals with feasibility of acquisition and management. Existing public lands in the planning area are assumed to have some degree of protection; however, planners intend to make recommendations for changes in management of these lands on the basis of the biological studies, so that they will serve better as true reserves for the biological elements concerned.

Many planning processes start out ambitiously but later stumble when encountering political and financial obstacles. In the case of the Coachella Valley MSHCP, much work remains to be done to complete the plan, so both the strengths and weaknesses of the plan are subject to considerable change. Identifying the strengths and weaknesses of the current effort was the main charge given to the conservation biologists who served as outside experts in the late-1996 workshop. Among the strong points noted by the reviewers is that the plan takes a broad, landscape focus and considers many different natural communities, all of which interact ecologically through shared species and natural processes. Not only does the multiple-community approach make sense ecologically, it should help provide better assurances to landowners. Unlike the case for some HCPs, participants in the MSHCP will not have to enter into new planning processes as rare species associated with habitats not covered by the plan are petitioned for listing under the federal or state ESAs.

Another strength of the plan noted by reviewers is its proposed use of GIS-based vegetation mapping and gap analysis. If successful, this will be the first HCP or NCCP that effectively applies a coarse filter strategy. Working against this, however, is the political reality that an argument for habitat protection is weaker when it is not explicitly tied to the survival of species listed or likely to be listed soon under the federal or state ESAs. When large amounts of land and money are at stake, the long-term advantages of proactive, ecosystem-level conservation may not measure up to the exigent argument of only saving that habitat known to be needed by listed or candidate species for which incidental take permits may be required. In addition, there are no quantitative criteria in the planning guidelines for determining what is "adequate" representation of natural communities in reserves. The reviewers also noted the desirability of preparing a map of historical vegetation for the planning region, for comparison with the map of current vegetation, so that the vegetation types ("endangered ecosystems") that have suffered the most severe losses can be highlighted. Preparing such a map appears to be feasible in this case.

Several aspects of the proposed Coachella Valley MSHCP are not consistent with the criteria for adequacy offered in this chapter. Weaknesses of the current planning effort that will hopefully be corrected as the plan evolves include a general lack of detail and scientific rigor in assessment guidelines; this is a problem common to many conservation plans. The planning approach is conceptually sound and reasonably comprehensive, but many of the questions and indicators are vague. Some subjectivity is unavoidable—and, in fact is desirable because it takes advantage of the considerable knowledge and expertise of the participants in a flexible and adaptive way—but quantitative criteria are desirable for ranking sites in terms of conservation value, as well as for monitoring. Measurable indicators need to be selected that correspond to each of the questions considered in selecting reserves (see chapter 6). For example, in assessing whether the food chain in an area is intact, the species (e.g., carnivores) and population sizes necessary for intactness must be specified. Indicators are also needed for level of fragmentation, connectivity, hydrological processes, and other questions on the site selection checklist. Many indicators of this type have been developed in other planning processes and are GIS-compatible (see Noss 1995 and chapter 6).

A general concern about the Coachella Valley MSHCP, which is common to many plans, has to do with the stated goal to obviate the need for listings of species under the federal and state ESAs. This goal would not be a concern—in fact, it would be laudable—if the plan not only were adequate for maintaining viable populations of all species that potentially deserve listing under the Acts, but that it actually would contribute to the recovery of all these species. As discussed elsewhere habitat-based conservation plans may be used, for better or worse, as substitutes for recovery plans. If this is the case here, then the MSHCP will have to qualify as a substitute by truly fostering recovery of multiple species. It will also have to contain a strong adaptive management capability to respond to changing conditions and new information.

Unfortunately, the current planning effort has not progressed to the stage of developing a monitoring program and guidelines for adaptive management, so it was impossible for the reviewers to assess these aspects of the plan. The only mention of plan modification in the current guidelines has to do with species not listed as Species of Concern that subsequently are recognized as needing protection. If the FWS or California Department of Fish and Game determines that any such species is not adequately protected by the plan, then it is proposed that an amendment to the MSHCP be prepared to afford such protection. We do not consider this limited amendment provision an adequate adaptive management program, because it includes no requirements for monitoring and neglects many other management questions that require consideration and adaptability.

As noted above, the MSHCP, like virtually all conservation plans, currently lacks sufficient funding for thorough biological surveys, land acquisition, monitoring, and other essential components of a scientifically defensible plan. This is the chief impediment to further progress. The planning workshop in late 1996 identified a number of research needs and also concluded that the implementation of the plan should be phased in, with most urgent attention given to blow-sand habitats (including dunes, sand sources, and wind corridors) and severely depleted natural communities, such as alkali sink scrub. Funding, we hope, will become available to set the plan in motion.

6

A FRAMEWORK AND GUIDELINES FOR HABITAT CONSERVATION

In this chapter we return to the lessons learned through the successes and failures of conservation plans over the years, combining them with new insights we see emerging from the theory and practice of conservation biology. Although conservation biology is still a young science, and potential success stories have not continued long enough to be entirely convincing, we consider the guidelines offered here robust enough to be applied now. Waiting for perfect, abundantly corroborated guidelines before proceeding with conservation would be foolish, as much biodiversity would be lost in the interim. Our guidelines are mostly general, because the diversity of the natural world makes it perilous to assume that a specific conservation practice that worked in one case will work in all others. As noted in chapter 4, conservation biology is a science of case studies (see Shrader-Frechette and McCoy 1993), and careless extrapolation is dangerous. Nevertheless, failure to apply defensible principles of conservation biology to real-life conservation work is even more dangerous.

The framework we offer is inherently adaptive and designed to be revised and improved as new information and techniques emerge and as theories are refined. It is flexible enough to be tailored to specific situations. It is intended to provide a basis for reconciling species-level and ecosystem-level conservation. The alternative to applying innovative theories and technologies and learning from lessons elsewhere is to plod along with the expensive, idiosyncratic, species-by-species, site-by-site, reactive approach that has failed almost everywhere. In conservation work, failure ultimately means extinction.

Below we review what we feel are several necessary components of a habitat-based *(in situ)* conservation plan: scientific evaluation of alternatives, spatially explicit assessment of population status or viability, reserve selection and design, sustenance of ecological and evolutionary processes, adaptive management (including use of ecological indicators for monitoring), and linkage of disparate planning efforts into a unified approach. These components are not sequential; that is, one should not wait until the last stages of conservation planning to develop ecological indicators and unite planning efforts. Rather, these components must be pursued simultaneously from the beginning of each plan.

Scientific Methods and
the Evaluation of Alternatives

Conservation planning must rest on a secure scientific foundation. Science operates to a large degree by evaluating alternative hypotheses. Conservation science operates, in part, by comparing alternative conservation plans, reserve designs, or management plans, then selecting those that accomplish stated objectives and discarding (at least temporarily) those that do not. Thus, conservation planning has many parallels to the National Environmental Policy Act (NEPA) process. In federal projects subject to NEPA review, agencies must prepare environmental impact statements (EISs) that consider the effects of a range of alternative courses of action on the environment; these EISs are open to public review. After review, draft EISs are revised into final EISs, with a preferred alternative and record of decision. Modern conservation planning is a similar process—and in fact much of it is subject to NEPA review—but in our opinion requires an involvement of scientists and scientific methods far beyond that typical for NEPA review, which is often criticized as purely procedural. Conservation planning, to be scientifically defensible, must employ scientific methods and be open to revision based on independent peer review. Further, as noted earlier (see box 3.1), it is imperative that outside scientific review and consultation begin early in the planning process.

In evaluating the potential effects of alternative plans or other courses of action, some of the most crucial decisions have to do with the methods to be employed in the analysis. Hypothesis testing is the best-accepted approach to scientific investigation today. Map-based conservation plans provide hypotheses that can be tested (Murphy and Noon 1992; see box 6.1). Using the interagency planning exercise for the northern spotted owl as an example (Thomas et al. 1990),

Murphy and Noon (1992) noted that statistical analyses of empirical data, predictions from ecological theory and population dynamics models, and inferences drawn from studies of related species can be used to test properties of a preliminary mapped reserve system. Through an iterative process of testing maps against data and model predictions, preliminary maps can be refined into a scientifically defensible network of reserves and other zones. Iterative map-based planning is now widely accepted as a legitimate scientific approach to solving conservation problems (see, for example, appendix 5B). With the availability of GIS (see box 6.2), alternative reserve designs can be displayed and conveniently tested against each other and against null models (i.e., no reserves or alternately no development) to predict different effects on biodiversity.

The question of what constitutes defensible scientific methodology in conservation planning needs some discussion here. Although we expect conservation science to be rigorous and defensible, we emphasize that standards for rigor and defensibility in conservation planning—and in much of the field research that feeds information to the planning process—should not be based on a laboratory-science model. Conservation planning exercises, even when designed to test and assess alternative hypotheses, rarely lend themselves to the kind of strictly replicated experimentation in which hypotheses can be falsified with high confidence using inferential statistics. As reviewed by Shrader-Frechette and McCoy (1993:81), using conventional hypothetico-deductive methods in ecology—especially field ecology—is fraught with problems because, among other things, "It is difficult to construct uncontroversial null models to test hypotheses; and . . . value judgments . . . often determine the relationship between evidence and theory." These difficulties do not mean we should discard theory or forget about testing hypotheses. "Just because facts are value-laden, this does not mean that there is no sufficient reason for accepting one theory over another. One theory may have more explanatory or predictive power, or unify more facts" (Shrader-Frechette and McCoy 1993:101). Hence, we are not sympathetic to apologies for sloppy science based on appeals to the inherent difficulties of testing hypotheses in conservation biology.

Those who seek to assess the rigor of science in a conservation plan must understand the nature of applied science and be willing to reconsider the narrow version of the scientific method they learned in school. Although most of the whole-organism and ecological sciences

Box 6.1. Hypothesis Testing and Reserve Design for Northern Spotted Owls

Given the charge by Congress in 1989 to produce a "scientifically credible conservation plan for the northern spotted owl" an interagency team headed by Forest Service biologist Jack Ward Thomas was granted a first-of-its-kind opportunity to bring science to bear in conservation planning. The Interagency Spotted Owl Scientific Committee took that mandate and assumed that their product would have to be scientifically defensible as well. The explosive political situation in the Pacific Northwest guaranteed a lawsuit on delivery of the plan, if not by the timber industry, then by the environmental community—and probably by both. With its day in court inevitable, the spotted owl plan would have to be logical and rigorous, and its reasoning linear and repeatable.

The team agreed that for the plan to meet those criteria, scientific methods would have to be followed. In essence the plan would have to be structured like a scientific experiment, with tests of hypotheses using the best available data. But the process of falsifying hypotheses demands a prediction of the result. So, what form would those hypotheses take? A completed habitat-based plan could have a number of quantifiable characteristics—an amount of habitat, a number of patches, distances that separate them—all mappable attributes. The team surmised that the hypothesis must, in fact, be a map. The ensuing scientific enterprise provided a model for the application of biological information in reserve design efforts for imperiled species.

Three initial hypotheses were tested with data in an attempt to validate the need for spotted owl conservation in the Pacific Northwest: (1) that populations of owls were actually declining, (2) that they are in fact habitat specialists, requiring late-successional forests for population persistence, and (3) that those forest habitats were diminishing in extent, exacerbating the threat to the species. Accumulated evidence indeed suggested that populations of this territorial species were declining and that the declines were best explained by losses of mature and old-growth forest habitat.

Results from tests of these fundamental hypotheses justified a conservation program designed to arrest and reverse the decline of the owl. To meet that goal, the committee adopted a map-based approach that would determine just how much habitat was needed by the owl and how that habitat should be distributed, including the size, shape, and spacing of habitat patches. The committee was also interested in identifying characteristics of the landscape matrix in which

Box 6.1. Hypothesis Testing and Reserve Design 159

a mature forest reserve system would be embedded. Toward the goal of maintaining a stable population of owls that would be widely distributed throughout their historic range, a reserve design portrayed as a map and its quantifiable properties was subjected to tests with empirical data and theoretical predictions. Four map layers were used: historic and current owl distributions, historic and current habitat distribution, known owl locations, and land ownership. Those layers when overlaid provided a preliminary reserve design, one that represented the maximum size and number of habitat patches that would support the owl.

The best biological solution was to designate all mapped habitat as part of a reserve system, but that solution was universally viewed as neither politically nor economically feasible—nor for that matter legally possible because of the multiple-use mandate under which the public lands of the United States are managed. Instead, a subset of actual or potential habitat patches that could sustain the owl was sought that would meet the conservation objective of a 95 percent likelihood that owls would survive for at least 100 years. This process of refining a map of habitat patches into an institutionally acceptable reserve system used empirical data from studies of the owl (along with data from similar species), predictions from simulation models, and more general ecological theory. In an iterative fashion, through a hypothesis testing scheme that allowed the scientists to retain or reject test results, a strategy emerged that provided the owl a network of forested habitat patches. Using four critical assumptions—that 35 percent of the forested landscape would exist within potentially conserved areas, that 60 percent of the patches would provide potential nesting habitat for the species, that owls in each patch would be within dispersal distance of at least one other patch, and that female owls would have no difficulty finding mates—the testing process suggested a reserve of patches each of sufficient size to provide for 20 pairs of territorial owls, with those patches located within 19 km of each other (the recorded dispersal distance of two-thirds of emigrating juveniles).

This repeatable exercise had one most important outcome—it allowed the scientists to defend their logic and product successfully in subsequent lawsuits brought by the timber industry. Lawyers for the industry sought to undermine the plan's credibility by identifying the weakest data or empirical generalizations. But the spotted owl plan was not really directly built from those data—it was actually tested with those data. Instead of being infinitely assailable and as weak as the weakest data used in its structure, born from hypothesis testing it was as strong as the strongest test of its integrity. Having exhausted the best available scientific evidence in its development, the committee produced a plan that was truly "scientifically credible."

Box 6.2. GIS As a Conservation Tool

James R. Strittholt (guest author) and Reed Noss

Maps have always played a significant role in conservation, but since the arrival of the computer age mapping has undergone a dramatic transformation. The conservation community has made a substantial shift in recent years away from traditional, manual mapping in favor of rapidly expanding computer-based technologies, especially GIS (geographic information systems) and remote sensing. Today, it is difficult to find conservation planning anywhere, at any scale, not incorporating GIS to some degree. Discussions about data layers, overlays, projections, data transfer, and so on have become commonplace. Computer mapping technologies and our ability to use them are still in an adolescent stage but are maturing rapidly. GIS has changed the way we look at conservation, and it has deeply penetrated conservation planning by academia, government agencies, nonprofit organizations, and others. Why this major shift toward computer mapping, and what does the future hold?

GIS has been described as having two fundamental utilities: a descriptive function and a prescriptive function (Tomlin and Johnston 1990). Descriptive mapping in modern GIS largely involves the computerization of traditional cartography, whose primary purpose is to characterize, or *describe*, the Earth in meaningful ways. Mapping such features as vegetation, soils, contours, and land ownership fall under this category. By transferring data and information from analog (most commonly, paper maps) to digital (or computer readable) form, a wide range of map products can be readily generated and updated with great speed and creativity. Although descriptive mapping allows ready calculation of acreage and other spatial statistics, and leads to some new insights as new combinations of geographically referenced data are blended together, it is really little more than high-tech cartography.

Prescriptive mapping, on the other hand, introduces something new and exciting to conservation. It builds upon descriptive activities by developing geographically based models of natural and human altered systems that can help planners understand these ecosystems better and plan for their protection and management. With the rapid advances in technology in recent years, GIS has become a primary tool for the maturing field of landscape ecology (see Haines-Young et al. 1993) and is well-suited to model regional landscapes at multiple scales. GIS-based models have been developed for such things as (1) hydrologic regimes (Maidment 1996), (2) forest succession and gap dynamics (Acevedo et al. 1996), (3) fire prediction and behavior, (4) habitat suitability (Mladenoff et al.

Box 6.2. GIS As a Conservation Tool 161

1995), and (5) ecosystem classifications (Host et al. 1996). Not only do these techniques allow scientists and planners to examine natural phenomena in new ways, they also afford the opportunity to produce, and to some degree evaluate, any number of alternative prescriptions, emphasizing different policies or management actions before any action is ever taken on the ground. For example, conservation priority setting is one popular area of scenario building (Bedward et al. 1992; Pressey et al. 1993; Scott et al. 1993; Kiester et al. 1996), as is evaluation of alternative reserve designs (Strittholt and Boerner 1995). GIS is also an ideal tool for implementing adaptive management. The capacity to assess multiple scenarios for problem solving and to organize ecological monitoring programs is among the more important technical contributions GIS is making to conservation.

Although we have made it a point to emphasize the differences between descriptive and prescriptive mapping, most GIS-based conservation projects contain aspects of both. When faced with a particular conservation problem, the logical place to begin is assembling the necessary data needed to describe current conditions—a descriptive mapping exercise. This is no simple matter under any circumstances, but when GIS is used this task can seem overwhelming at first. Although the benefits of incorporating GIS into planning are many, the time, money, and expertise required to take full advantage of the technology can be staggering. Just a few years ago it was not unusual to have as much as 80 percent of a project's budget go into building the initial GIS database. This burden discouraged many planners from adopting computer mapping technologies. Today, as a reflection of the commitment by government and many in the private sector to improve GIS technology and provide much of the data that feeds it, many more datasets are readily available and affordable to a growing number of users. This is particularly good news to nonprofit conservation groups, who regularly work under very tight budgets and time lines. Having more geographically referenced data readily accessible and available at low cost allows for more project resources to go toward actual conservation planning (i.e., prescriptive mapping), rather than toward the more menial task of constructing GIS data layers.

Despite the benefits GIS is beginning to provide to conservation, there are some drawbacks. One serious danger in adopting high technology such as GIS is the temptation to fund the latest gadgetry at the expense of existing programs. In conservation this may translate into cutting budgets for biological research in order to pay for expensive GIS equipment, software, and skilled technicians. Another danger is that the maps and models are so compelling that those who view them may assume they are based on extensive field observations and other data. Many are, but it has become increasingly easy to produce professional-looking results based upon very poor data and little or no real science. Furthermore, most GIS technicians are not qualified to conduct the types of analyses needed for defensible conservation planning without much interaction and support from scientists. Adopting GIS as a conservation tool should in no way justify reducing the input from scientists or the collection of basic field data. Rather, now that planners have a technology that can integrate vast amounts of disparate data and build sophisticated models, it is in their best interest to make the most of these newly emerging opportunities by emphasizing the quality of information that is fed into the computer. Field work to collect data on biological elements of interest and to validate (ground-truth) maps is needed more than ever. Unfortunately, educational institutions today emphasize course work in high-tech fields such as GIS and remote sensing at the expense of ornithology, entomology, taxonomy of vascular plants, and other natural history studies that make competent field work possible (Noss 1996c).

If the potential pitfalls can be avoided, the use of GIS and other computer mapping technologies will continue to grow, and the contribution GIS will make to conservation will be enormous. When used to its fullest potential GIS is an unsurpassed tool for integration and synthesis of the information required to address conservation problems. Successful conservation plans and other solutions to conservation problems are not going to come from the results of any particular study. Rather, progress will in most cases be made incrementally and will come from the thoughtful analysis and consideration of many kinds of information. GIS can contribute to the democracy of conservation planning by bringing everyone to the table—or computer screen—to see the world in new ways and, we hope, to plan for the protection and intelligent management of ecosystems.

pose difficulties for researchers trying to test hypotheses, the "applied" areas of ecology and conservation biology are more difficult than the "basic research" areas to align to a strict scientific method. Applied science seeks to "clarify the feasibility and consequences of alternative courses of action in particular contexts" (Brunner and Clark 1997:52) and to solve particular problems (e.g., the endangerment of species and ecosystems). Because many of the important variables and the relationships between them are unknown in applied science, a certain amount of trial-and-error learning is inevitable.

"Attempts to restrict 'science' to the hard science paradigm alone would inhibit our ability to learn from experience beyond the closed context of the laboratory" (Brunner and Clark 1997:53).

In conservation biology and applied science in general, there is no one scientific method. Rather, the researcher must be creative and use a mix of induction and deduction, analysis and synthesis, experimentation and practice-based learning. Scientists are trained mostly how to apply deductive methods of analysis, yet some of the most momentous contributions in the history of science have been based on creative thinking, careful observation, and syntheses of available information (e.g., Darwin 1859 and MacArthur and Wilson 1967). Thus, hypothesis testing in conservation science must be considered more liberally than in the conventional hypothetico-deductive model. Development of the principles of ecology and conservation biology is fundamentally an inductive, synthetic process in that one reasons from the particular to the general. Induction can be precarious because of the idiosyncratic nature of specific cases—anecdotes by themselves cannot be trusted. With enough anecdotes and case studies, however, real patterns emerge. "To do science is to search for repeated patterns, not simply to accumulate facts" (MacArthur 1972:1). These patterns are empirical generalizations, and when they are logical and widely agreed upon by a community of peers, they become guiding principles. We do not want to imply that guiding principles should not be challenged; in fact, they should be tested and scrutinized continually. But in the absence of case-specific information obtained from years of intensive research, well-established principles based on a collection of case studies provide a solid foundation for conservation planning.

With the principles of conservation biology reviewed in chapter 4 as a starting point, one can develop a series of questions and hypotheses pertinent to any case study at hand. Using the same principles combined with map-based analyses, one can compare plan alternatives in terms of their ability to attain conservation goals. Questions might be developed for several spatial scales of investigation (i.e., several levels of resolution) around several main topics known to be ecologically important for the species, ecosystem, or region concerned. Some examples of research questions pertinent to conservation of coastal sage scrub in southern California are included in box 6.3. Each of these questions could generate a number of specific, testable hypotheses. In practice, time and funding will always be

Box 6.3.

Examples of research questions for conservation planning for coastal sage scrub. A variety of specific hypotheses corresponding to each of these questions could be developed and tested.

Questions about Distribution

1. What is the general distribution of coastal sage scrub in relation to other habitat types within the region? (Both historic and present distribution should be mapped.)

2. How does the present distribution of coastal sage scrub correlate with physical and geographic variables such as physiography, elevation, topography, bedrock geology, soils, precipitation, distance from coast, and watersheds? Which of these environmental variables have the strongest effect on presence/absence and abundance of coastal sage scrub?

3. How does the present distribution of coastal sage scrub compare with the natural or historic distribution? (The extent of decline in acreage should be quantified, in addition to qualitative changes in the structure, composition, or function of the community.)

4. How might the present distribution of coastal sage scrub be partitioned into subregions for practical planning purposes? (The key issue for partitioning is "independence," which might be defined by a low level of biological interaction. Natural and anthropogenic factors have created a clumped distribution of coastal sage scrub patches that might define subregions that are relatively independent. However, political factors such as county boundaries are also relevant.)

5. Where are the large, relatively contiguous blocks or clusters of coastal sage scrub that might serve as core areas in a regional reserve network?

6. What is the distribution of particular rare species or groups of species across the region and in adjacent regions? (Are occurrences of these species correlated with one another or with environmental factors, or are they essentially random?)

7. What is the protection status of large blocks of coastal sage scrub in the region?

Box 6.3.   165

Questions About Species Trends and Interactions

1. Do population trends in different species associated with coastal sage scrub parallel one another, suggesting a common response to weather or other environmental conditions? Or do some species or groups increase at the expense of others, suggesting interspecific competition or divergent responses to the same environmental conditions?

2. Does cover of native shrubs and exotic grasses vary independently or correlate with each other, and how does this affect animal abundances?

3. Does the loss of cougar in a landscape lead to an increase in deer or change in the distribution of deer, and how does this affect vegetation?

Questions About Fragmentation and Connectivity

1. What species associated with coastal sage scrub are most vulnerable to fragmentation? (Some candidates are California gnatcatcher, cactus wren, orange-throated whiptail, cougar, coyote, mule deer, small mammals, and flightless invertebrates).

2. For those species known or suspected to be sensitive to fragmentation, what kinds of landscape features serve as absolute or partial barriers to movement?

3. Does a particular swath of habitat function as a corridor? For which species? Which landscapes might serve as major linkages between large blocks of coastal sage scrub?

4. How wide and long should a corridor be for each particular fragmentation-sensitive species?

Questions About Disturbance and Succession (e.g., Fire Ecology)

1. Does coastal sage scrub need to burn in order to regenerate and exhibit the full natural sequence of successional stages? (If it does, species richness should decline over time with lack of fire, or species composition should shift to species characteristic of other communities. Structural changes in vegetation and changes in processes such as nutrient cycling might also be noted.)

2. How often does coastal sage scrub need to burn or otherwise have succession set back in order to maintain its biodiversity?

3. How does the seasonality of fire affect the response of the community and its component species? (Fires in late fall or early winter, before the onset of winter rains, are thought to promote rapid reestablishment of native cover, decrease the time available for weeds to disperse onto the site, and reduce root mortality of native shrubs. In contrast, spring burns may increase weed invasion.)

limited, so it is important that scientists arrange questions and hypotheses in priority order—based on conservation urgency—and not be biased by their special areas of expertise (including the interest in funding). For this reason it may be preferable that generalists (scientists with broad training and research interests) rather than narrow specialists be charged with establishing overall priorities, although in some cases a team of specialists might be able to accomplish this task. To avoid bias, political considerations should be excluded from the process of identifying research questions and priorities.

Evaluation of conservation planning alternatives, then, involves careful deliberation of their hypothesized effects on the species and ecosystems of concern. Independent peer reviewers of a draft plan should, among other things, assess whether the evaluation of alternatives in the planning process was sufficiently rigorous in its application of scientific methods and unbiased in its final evaluation. Again, reviewers should be brought in to comment on the scientific planning process well before a draft plan is completed, so that opportunities for revision or change of course are available.

Assessment of Population Status and Viability

Conservation planning deals with real populations in real landscapes, not the abstractions of theoretical ecology and population biology. Thus, the ability to analyze spatial data on the status and distribution of species and habitats, and how organisms move across the landscape, is crucial. Assessment of the population viability of imperiled species has become one of the central research areas of conservation biology. At the same time, new computerized technologies have greatly expanded the ability of scientists to analyze and display spatial data. One of the most important advances has been the linkage of GIS mapping (see box 6.2) and simulation modeling, as applied to population viability analysis (PVA).

The viability of target species is a central concern in habitat-based conservation planning, for good reasons (see chapter 5, box 5.1). Most early PVAs were highly simplistic and focused on single, undivided populations. Before long, however, the recognition that landscapes are heterogeneous and that many species are distributed as spatially subdivided populations (Fahrig and Merriam 1994) created a demand for PVA models that could analyze dispersal and interchange of individuals among the subpopulations of a patchy population or metapopulation. As summarized by Quintana-Ascencio and Menges (1996:1211) "metapopulation models analyze the regional persistence dynamics of groups of populations by addressing the distribution of extinction risks and rates of migration among patches." Especially useful are spatially explicit population models—that is, models that combine a population simulator with a landscape map that shows the spatial distribution of habitats and other features in the landscape (Dunning et al. 1995). Such models incorporate locations of habitat patches and individuals and allow one to assess how differences or changes in landscape pattern affect populations. For example, reserve designs can be modeled to predict the effects of alternative designs on the population viability of a target species. Applying these models to multiple target species would be quite challenging, and we know of no successful examples.

A review of the spatial and spatially explicit PVA models now available and how they have been applied to conservation planning is beyond our scope. We simply offer a few examples and caveats, followed by examples of approaches that can be used when the data or time necessary for a full PVA are not available. Conservation planning for the northern spotted owl is one of the first examples where PVA was used to make explicit policy and management recommendations (Lande 1988; Doak 1989; Thomas et al. 1990; Lamberson et al. 1992). A review of the spotted owl models by McKelvey et al. (1993) summarized some of the lessons learned. For example, models determined that population growth is highly sensitive to variation in adult survival rate and relatively insensitive to fecundity or preadult survival rate. As habitat fragmentation continues, however, the problem of successful dispersal has more influence on the likelihood of persistence. Some inferences may be drawn from these models: (1) in landscapes experiencing a high rate of habitat loss, both occupancy and demographic rates may underestimate the risk of extinction; (2) large clusters of territories spaced farther apart are more stable than smaller clusters closer together; (3) the shape of clusters is nearly as important as their size; and (4) in landscapes that vary in habitat quality, the presence of marginal (sink) habitat adjacent to a reserve may increase extinction rates (McKelvey et al. 1993).

Another enlightening example of spatially explicit analyses is the use of a simulation program, ALEX, to model the viability of the endangered Leadbeater's possum *(Gymnobelideus leadbeateri)* in a network of forest patches in Australia (Lindenmayer and Possingham 1996). The analysis indicated that the effects of wildfire and salvage logging in forests inhabited by the possum resulted in a high probability of extinction, and that the design of reserves should consider impacts of wildfires, the size and spatial arrangement of habitat patches, the ability of animals to move among reserves, and the population size at which demographic stochasticity and environmental variation become influential (Lindenmayer and Possingham 1996). All of this information has direct relevance to reserve design and forest management.

PVA sounds almost too good to be true, and, indeed, in too many cases analyses are conducted recklessly and the results accepted with little questioning. The major limitation of spatially explicit models and PVAs generally is that they are enormously data-hungry. That is, "the cost of increasing 'realism' is the large number of model parameters and assumptions, which are often difficult to estimate and verify, respectively" (Hanski et al. 1996). Often the major limiting factor in conservation planning is not the availability of adequate models, powerful computers, or accurate maps of the landscape. Rather, what is lacking are accurate census and demographic data on the species concerned and sufficient knowledge of life histories, dispersal capacities, and other aspects of autecology. For example, the Scientific Review Panel for the coastal sage scrub NCCP was unable to provide rigorous guidelines for the connectivity of reserve networks necessary for the persistence of a metapopulation of California gnatcatchers. Data on dispersal distances of gnatcatchers were rudimentary, and virtually no information was available on the kinds of habitats that gnatcatchers will disperse through and what constituted a dispersal barrier. Hence, early metapopulation models for the gnatcatcher (M. Gilpin, unpublished) were abstract and highly uncertain. A more recent metapopulation model for the gnatcatcher in Orange County (Akçakaya and At-wood 1997) was based on more (though still incomplete) data and was able to validate a habitat suitability model for the species and build a model that was stage-structured, stochastic, and spatially explicit. But still, estimates of persistence time and other important predictions were uncertain.

We caution that although conservation planning ideally should make use of the best available technologies for PVA, predictions based on these methods are only as good as the data available. Flawed predictions can damage the credibility of the science applied to planning. Unfortunately, the trend in biological science is increasing numbers of professionals with expertise in computer sciences and modeling and a declining cadre of field naturalists with the expertise and experience to collect reliable, basic data on species that can inform such models (Noss 1996c). Errors in the estimates of model parameters, such as habitat suitability, demographics, dispersal distances, and other factors, can lead to incorrect predictions. Ruckelshaus et al. (1997) examined a simple model of organisms dispersing through a fragmented landscape to see how errors in input parameters translated to errors in model predictions. As expected, in an error-free model dispersal success was higher when more of the landscape was composed of suitable habitat and in landscapes filled with more small patches rather than fewer large patches. Errors in estimates of mortality during dispersal, mobility, and landscape conditions led to flawed predictions. The dispersal-mortality errors had the most serious effect, yet data of this type are among the most difficult and time-consuming to collect. Prediction errors were greatest in landscapes with a lower percentage of suitable habitat, "precisely the type of habitat characterizing most species of conservation concern" (Ruckelshaus et al. 1997). Because spatially explicit PVA models are often more detailed than the available data, these authors recommend more empirical studies, as well as use of simpler models.

Despite such problems PVAs and similar models have considerable utility even in the absence of reliable data. Their usefulness in such cases is not for making specific predictions but for addressing general "What if?" questions about effects of alternative reserve designs or management practices, for generating testable predictions, and for determining the specific kinds of data needed to apply or validate the models. These may be the areas where PVA contributes most meaningfully to conservation planning. For example, sensitivity analyses can be undertaken on life history parameters to identify the processes or life stages that are most critical to population persistence and the variables for which more information is required before a reliable PVA can be performed (Possingham et al. 1993; McCarthy et al. 1995; Norton 1995). In the study of Leadbeater's possum, the authors concluded that "probably the most important outcome from the application of PVA is the ranking of various conservation strategies and the testing of the sensitivity of that ranking, rather than an accurate definition of a minimum viable population" (Lindenmayer and Possingham 1996). Similarly, Akçakaya and Atwood (1997) concluded that their gnatcatcher model was much more useful for comparing management options than for making absolute predictions about the persistence time of the population.

If conservation planners decide that PVA of one or more target species is desirable for a particular plan, the choice of which program to use could be pivotal. A variety of "canned" programs is available, and opinions vary widely among researchers as to which is best in any particular situation. If all programs gave similar predictions, the choice would not matter much. This may not be the case, however. In a test of four common programs using a grizzly bear data set, Mills et al. (1996) found that idiosyncrasies of input format for each program led to minor differences (3 percent) in intrinsic growth rate that produced major disparities in estimates of expected population size and extinction rate. Furthermore, when density dependence was added to the models, the programs produced widely varying predictions. Mills et al. (1996) recommend that unless the data strongly suggest a particular density-dependent function, at least one modeling scenario should lack density-dependence. Further, whenever possible, several PVA programs should be used for each case, along with a range of parameter estimates within each (Mills et al. 1996). Another important lesson from this study is that even with abundant field data, predictions from PVAs using different programs will diverge from each other—and from reality—simply because of the peculiarities of the programs.

We expect there will be relatively few conservation plans for which full-blown PVA analysis, using several models, is affordable in either time or money. Fortunately, shortcuts for viability analysis can be taken when the detailed biological data needed for complex simulation models are lacking. A very useful approach, called "species-centered environmental analysis," by James et al. (1997), begins by organizing prior knowledge about the factors that limit the population of interest (fig. 6.1). From there, additional studies can be conducted to evaluate alternative explanations for how environmental factors affect populations and how management might be changed to benefit the population. Accumulation of knowledge in this approach is incremental. Hanski et al. (1996) described a simple, spatially explicit incidence function model that, when applied to a metapopulation of an endangered butterfly (the Glanville fritillary, *Melitaea cinxia*), predicted patch occupancy well and allowed for quantitative predictions about metapopulation dynamics. In this model the probability of local extinction is determined by the size of the respective habitat patch, and the probability of colonization of an empty patch is determined by its isolation from the occupied patches and the sizes of these patches. The model does not require detailed information on demographic parameters, but uses "snapshot" presence/absence data from a collection of habitat patches to estimate biologically important metapopulation parameters (Hanski et al. 1996). The model assumes that the metapopulation from which the snapshot is taken is not far from a stochastic steady state, an assumption that might not be warranted for endangered species that are declining along with habitat destruction (Wahlberg et al. 1996).

Even presence/absence data are difficult to gather on species that are extremely rare. Wahlberg et al. (1996) were able to avoid this problem by estimating model parameters for another endangered butterfly *(Melitaea diamina)* from data collected on the well-studied and ecologically similar *Melitaea cinxia* (Hanski et al. 1996). The model predicted the occurrence of *M. diamina* reliably and confirmed that the metapopulation of that species was, in fact, close to a steady state and that metapopulation dynamics were dominated by the effects of patch area and isolation. Most important, from a conservation planning standpoint, the model helped identify patches and landscapes that are critical to the persistence of the species; destroying (by simulation) peripheral subpopulations on isolated patches had little effect, whereas destroying patches in more central positions caused the metapopulation to collapse (Wahlberg et al. 1996).

In another study Quintana-Ascencio and Menges (1996) applied the incidence function model to populations of 123 species of vascular plants and ground lichens in 89 patches of Florida scrub. The data were simple patch occupancy records for each species and information on patch size and isolation (using indices and GIS mapping) and fire

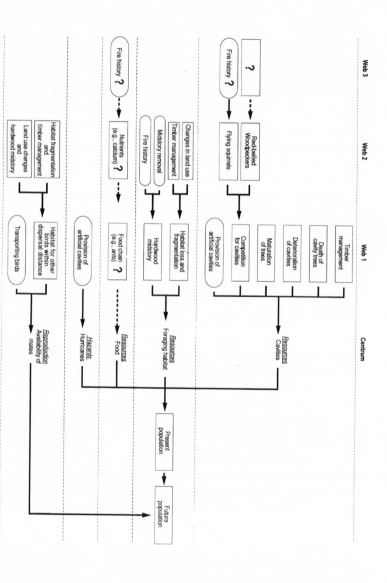

FIGURE 6.1. An envirogram for the red-cockaded woodpecker. The centrum includes environmental factors that, if their levels were changed, would be expected to affect the numbers of woodpeckers directly. The webs include progressively less direct influences, most of which, however, can be modified by management. From James et al. (1997), who modified the envirogram format from Andrewartha and Birch (1984).

history. Their results showed strong effects of patch size and patch isolation on herbaceous species closely associated with rosemary *(Ceratiola ericoides)* scrub and suggested that fire suppression and habitat destruction may decrease persistence probabilities for these species, many of which are endangered (Quintana-Ascencio and Menges 1996). Thus, this simple model was able to provide crucial information for management of rosemary scrub.

Other streamlined approaches to assessing the status and viability of populations and metapopulations are available. Lacking data on absolute population sizes for tigers, Wikramanayake et al. (in press) used ten-year demographic trends to evaluate the conservation potential of tiger populations. Felids are notoriously difficult to census and results are unreliable without substantial effort; thus, obtaining enough data for a valid PVA for the tiger was impossible. The trend estimates proved very useful, however. Wikramanayake et al. (in press) concluded that "These general trend estimates are more easily obtained from field staff and people familiar with local conditions than absolute numbers of animals, and very likely reflect the population status of tigers with as much precision as would the available data on population numbers and PVAs. Trends also reflect trajectories, which are important for conservation planning and management." Similarly, the IUCN criteria for Red Lists (IUCN Species Survival Commission 1994) assign species to categories based on any of several lines of evidence: population reduction, areal extent and pattern of occurrence, population size, or a quantitative viability analysis (box 6.4). We believe that these kinds of estimates, especially trends, will provide much of the information needed by planners and managers who must select target species, assess their status or viability, and judge the effects of management but, for various reasons, are unable to complete PVAs.

Nearly all assessments of viability for conservation planning have concentrated on single species, or if several species were included, their viability was considered separately. But many species have strong interactions—positive as well as negative—with other species. One of the best-known positive relationships, or mutualisms, is that between figs and fig wasps. In this obligate mutualism fig trees depend on the wasps to pollinate them, and the wasps can persist only when there is an adequate population of figs. Anstett et al. (1997) combined field observations, experiments, and simulation modeling

Box 6.4.
IUCN Red List Criteria for
Critically Endangered species

Criteria for Endangered and Vulnerable species are the same, but with different numbers. From IUCN Species Survival Commission (1994).

Critically Endangered

A taxon is Critically Endangered (CR) when it is facing an extremely high risk of extinction in the wild in the immediate future, as defined by any of the following criteria (A to E):

    A. Population reduction in the form of either of the following:

        1. An observed, estimated, inferred, or suspected reduction of at least 80 percent over the past ten years or three generations, whichever is the longer, based on (and specifying) any of the following:

            a. direct observation

            b. an index of abundance appropriate for the taxon

            c. a decline in area of occupancy, extent of occurrence, and/or quality of habitat

            d. actual or potential levels of exploitation

            e. the effects of introduced taxa, hybridization, pathogens, pollutants, competitors, or parasites.

        2. A reduction of at least 80 percent, projected or suspected to be met within the next ten years or three generations, whichever is the longer, based on (and specifying) any of (b), (c), (d), or (e) above.

    B. Extent of occurrence estimated to be less than 100 km^2 or area of occupancy estimated to be less than 10 km^2, and estimates indicating any two of the following:

to study the persistence of this two-species interaction. Their results showed that rare fig species and their pollinators are the most vulnerable to habitat fragmentation and that vulnerability is highest in species or sites where flowering is highly seasonal. When two or more target species for a conservation plan are suspected to interact

BOX 6.4. IUCN RED LIST CRITERIA 175

1. Severely fragmented or known to exist at only a single location.

2. Continuing decline, observed, inferred, or projected, in any of the following:

    a. extent of occurrence

    b. area of occupancy

    c. area, extent, and/or quality of habitat

    d. number of locations or subpopulations

    e. number of mature individuals

3. Extreme fluctuations in any of the following:

    a. extent of occurrence

    b. area of occupancy

    c. number of locations or subpopulations

    d. number of mature individuals

C. Population estimated to number less than 250 mature individuals and either:

    1. An estimated continuing decline of at least 25 percent within three years or one generation, whichever is longer, or

    2. A continuing decline, observed, projected, or inferred, in numbers of mature individuals and population structure in the form of either:

        a. severely fragmented (i.e., no subpopulation estimated to contain more than 50 mature individuals), or

        b. all individuals are in a single subpopulation.

D. Population estimated to number less than 50 mature individuals.

E. Quantitative analysis showing the probability of extinction in the wild is at least 50 percent within ten years or three generations, whichever is the longer.

strongly, studies of their connected population dynamics and viability should be considered.

We urge that conservation planners make every effort to acquire the basic field data necessary for reliable assessments of the status and trends of target species. At a minimum this information, collected over time, will be useful in designing and implementing adaptive management strategies. In our experience planners have put too little emphasis on field surveys and autecological research of target species and their interactions. Without knowledge of the life histories and ecological requirements of target species—or even, in many cases, information on the presence or absence of target species in habitat patches—conservation plans will be vague and likely to require substantial modification over time. Such plan revisions will be both expensive and politically sensitive, particularly when assurances have been made to landowners and other private participants. It makes much more sense to address these issues early in the planning process. Nevertheless, even in the absence of detailed autecological and demographic data, PVAs and other population assessments can be instructive in identifying the life history parameters most sensitive to disturbance and the factors most likely to threaten the species. Moreover, they allow alternative reserve designs or management options to be compared in terms of their potential effectiveness. For an example of how useful these types of data can be in designing a conservation plan, see the case study described in appendix 5B.

Reserve Selection and Design

Virtually all conservation plans have proposed some kind of reserves, ranging from small set-asides in the midst of development to expansive networks of protected areas on a regional scale. Recently, however, the reserve strategy has been criticized on several fronts. Most of the responsible critics of protected areas offer some model of sustainable development and resource use as an alternative. These proposals include everything from extractive reserves and community-based development projects in developing countries to New Forestry in the Pacific Northwest and "holistic resource management" on rangelands. But optimistic assumptions about the ability of humans to interact harmoniously with Nature in a "landscape without lines" (i.e., no reserves or zoning) are naive. Most experiments to date in sustainable development, sustainable use, multiple use, and ecosystem management have been failures (Redford and Sanderson 1992; Ludwig et al. 1993; Robinson 1993; Irvine 1994; Noss and Coooperrider 1994, Stanley 1995). This does not mean we should stop experimenting with new resource management or development practices. Such experiments are urgently needed so that we can determine which human uses and levels of use are compatible with conservation objectives and which are not. But it is unwise to experiment with every remaining natural area, as these areas become fewer and fewer. A valid experimental design includes control areas or reference sites that can be compared to manipulated areas and offer lessons about how ecosystems function relatively free of human influence (Leopold 1941).

The current consensus among biologists, as we interpret it, is that protected areas are necessary but not sufficient to meet conservation objectives (Robinson 1993; Meffe and Carroll 1994; Noss and Cooperrider 1994; Callicott and Mumford 1997). A revised biosphere reserve model of interconnected reserves enveloped in well-managed, multiple-use buffer zones or landscape matrix may offer the best hope of maintaining biodiversity over the long term, as well as meeting human needs (Noss 1992, 1996; Noss and Cooperrider 1994). This model is, in many cases, what regional conservation plans attempt to achieve, except in urban landscapes the role of buffer zones is often questionable. In regions where expansive reserve networks cannot be developed, smaller protected areas potentially play the role of safeguarding some of the species and habitats most sensitive to human activities, in addition to their values as benchmarks and natural laboratories. If the reserves are too small in the context of the surrounding land-use—and/or if adequate buffers are lacking—these values will be compromised by edge effects and the inability of the reserves to support even temporary populations of species with large home ranges or naturally low densities.

In chapter 4 we reviewed some principles of conservation biology that apply specifically to reserve design. Briefly, reserves should be large, multiple, and functionally connected. These principles are widely agreed on; thus, map-based hypothesis testing does not begin in a vacuum, but starts with a design (or better yet, a series of alternative designs) that conforms to these principles. Before one goes about designing reserves, however, one must identify the sites in the landscape that are most important to protect; this is the process often referred to as reserve selection or identification (Noss and Cooperrider 1994). Sometimes, core reserves are easy to identify in a fragmented landscape—they are the largest blocks of relatively undisturbed habitat that remain. More often, reserve selection requires an iterative process of site evaluation. Once the crucial sites are identified, then a reserve system can be designed around them. The process of reserve selection and design was recently reviewed by Noss and Cooperrider (1994) and Noss (1995, 1996b) and need not be repeated in detail here. One useful approach in many cases is for the process to proceed along three major and parallel tracks: mapping of rare species occurrences and other special elements and sites, representation (gap) analysis, and spatially explicit population assessments for area-dependent or otherwise highly vulnerable species. These three tracks converge when information from each is combined in map form and subjected to the rules of reserve design (fig. 6.2; Noss 1996b). Thus, sites identified as necessary to protect target species' populations and natural communities will be the core areas that, whenever possible, should be secured, buffered, and connected (if they were naturally so prior to human disturbance) to form a functional network of reserves across the planning region.

Reserve designs will normally be difficult to implement all at once. Hence, a sequential approach is often favored, where private lands are gradually acquired and added to the public land base or are protected by conservation easements or other voluntary agreements, or where public lands are taken out of resource production and designated as wilderness, research natural areas, or other reserves. This process, given the best of circumstances, may take decades or more. Therefore, conservation evaluation procedures should identify those sites that must be protected immediately in order to prevent losses of biodiversity and distinguish them from other sites that are necessary for long-term viability of species or ecosystems but face less urgent threats and need not be protected right away (Noss 1996b).

Sites needing urgent protection may include the only known populations of endemic species or the highest-quality examples of particular natural communities (by definition, these sites are irreplaceable), sites that provide refugia or source populations of target species, or other sites of undocumented but suspected high biological value that are in immediate danger of destruction. A spatially explicit population viability analysis, based on adequate demographic data, may be necessary to identify source populations unambiguously—and can be used to estimate time limits for land protection—but information on population trends can also be useful in identifying sites of high value to particular species (see the preceding section). Some of the sites targeted for later acquisition will inevitably be lost to development in regions experiencing rapid human population growth or resource extraction, so it is important to identify alternative reserve sites and

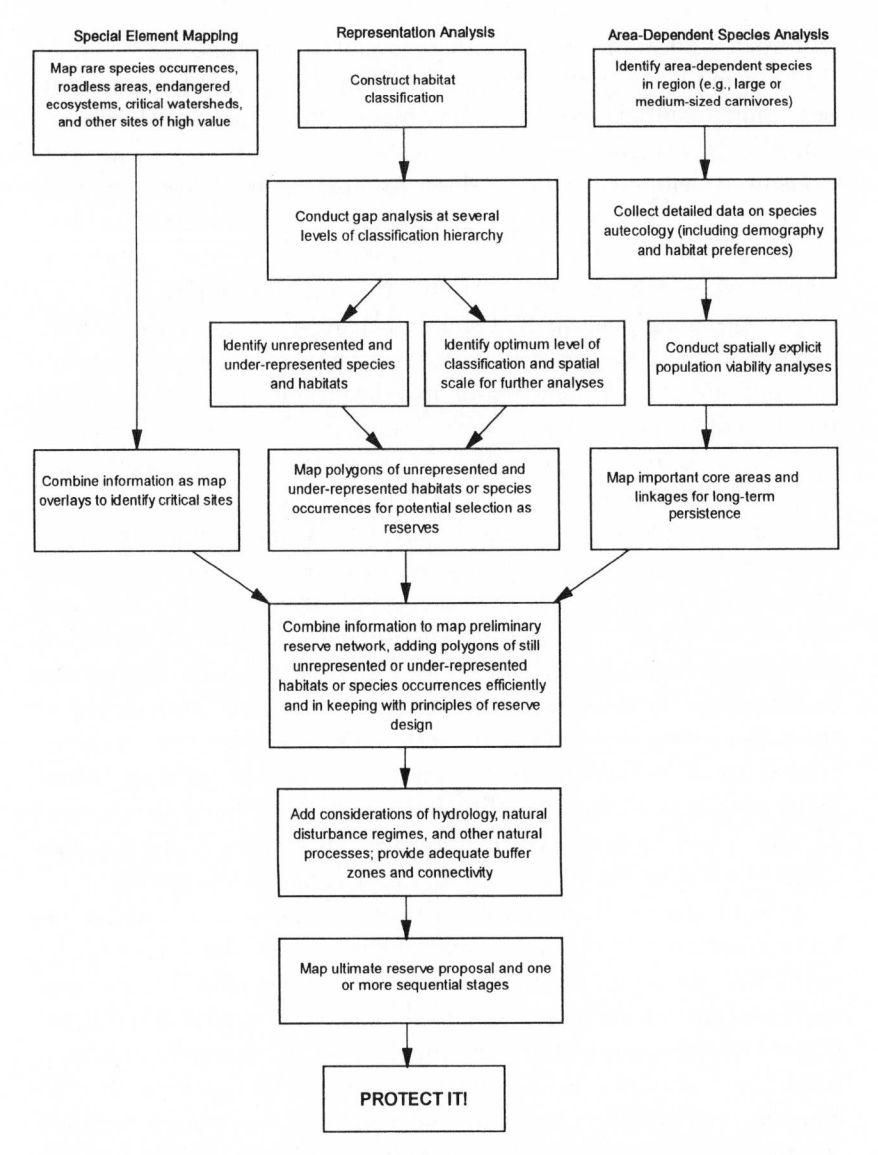

Special Element Mapping

Map rare species occurrences, roadless areas, endangered ecosystems, critical watersheds, and other sites of high value

Representation Analysis

Construct habitat classification

Conduct gap analysis at several levels of classification hierarchy

Identify unrepresented and under-represented species and habitats

Identify optimum level of classification and spatial scale for further analyses

Area-Dependent Species Analysis

Identify area-dependent species in region (e.g., large or medium-sized carnivores)

Collect detailed data on species autecology (including demography and habitat preferences)

Conduct spatially explicit population viability analyses

Combine information as map overlays to identify critical sites

Map polygons of unrepresented and under-represented habitats or species occurrences for potential selection as reserves

Map important core areas and linkages for long-term persistence

Combine information to map preliminary reserve network, adding polygons of still unrepresented or under-represented habitats or species occurrences efficiently and in keeping with principles of reserve design

Add considerations of hydrology, natural disturbance regimes, and other natural processes; provide adequate buffer zones and connectivity

Map ultimate reserve proposal and one or more sequential stages

PROTECT IT!

FIGURE 6.2. Three tracks of reserve selection and design, which converge in a defensible reserve system. These three tracks have rarely been combined in practice, as researchers usually pursue their specialized interests independently. Adapted from Noss (1996b).

minimize allocation of land to irreversible land uses. For example, grazing and logging often do not result in irreversible losses of conservation potential (exceptions may include loss of primary forest and rangelands transformed to exotic, annual grasslands), whereas housing development usually does. An abstract model of sequential reserve-network development in a managed forest landscape is illustrated in figure 6.3 (DellaSala et al. 1996) and a potential, partially completed sequence for Florida in figure 6.4 (Noss 1996b).

No one should assume that simply placing a substantial area in reserves will fulfill the goals of a conservation plan. What is a "substantial" amount of land in one region may be far too little in another. The question of "How much is enough?" was considered by Noss (1996b). Briefly, the amount of protected land required to meet conservation goals will depend on many factors, including the specific objectives of the plan, the physical and biotic heterogeneity of the planning region (regions that are more heterogeneous or have higher endemism will require more protected area), the area requirements of target taxa, and the area necessary for natural disturbances and other ecological processes to function normally (Noss 1996b). In highly fragmented and degraded landscapes, conservation goals as we have identified them often may be achieved by setting aside less land and concentrating more on habitat restoration and active management to enhance and sustain habitat quality. In some of these landscapes, unfortunately, the land available for conservation—even if protected in its entirely—may be too little to attain many conservation goals.

At least equal in importance to the overall amount of land in a reserve system are the configuration of the system, the nature of the surrounding landscape, and human uses. Wilcove (1985) found that birds nesting in woodlots surrounded by suburbs experienced higher rates of nest predation than those in woodlots surrounded by agricultural land. Similarly, Friesen et al. (1995) found the diversity and abundance of songbirds in forest blocks surrounded by suburbs much lower than in forests with few or no nearby houses, likely because populations of house cats, squirrels, and other nest predators are higher in suburbs. As another example, Staten Island, New York, which has 10 percent of its area protected (more than most landscapes in the United States, but mostly within a central greenbelt) lost 40 percent of its native plant species and 53 percent of its rare and endangered plants between 1879 and 1991, while exotic plants increased from 19 percent to >33 percent of the flora (Robinson et al. 1994). The

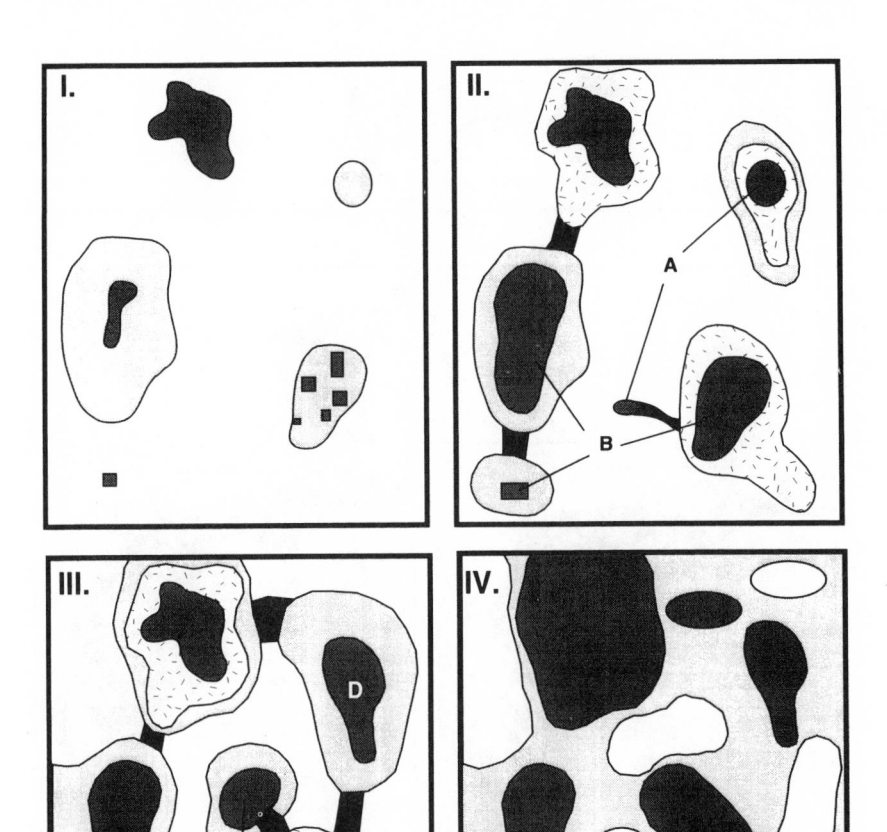

FIGURE 6.3. Conservation planning and implementation stages for a managed forest landscape, using a core, corridor, and buffer model in the early stages but resulting in a wildland matrix. Plate I depicts current conditions, Plate II the first phase of increased protection (10–15 years), Plate III the second phase (50 years), and Plate IV the final goal. Dark gray = protected areas; light gray = unprotected natural areas (I) or buffer zones (II–IV); confetti pattern = restoration zones; black = corridors; white = intensively managed forest. Letters A–E indicate progressively expanding protected areas. From DellaSala et al. (1996).

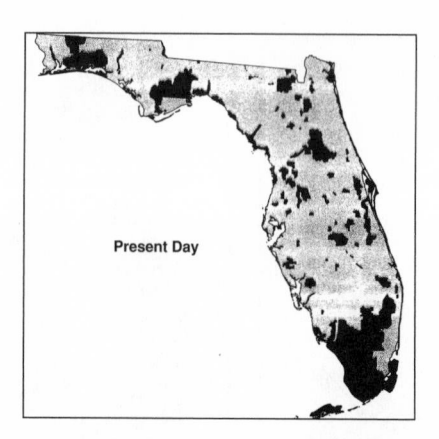

FIGURE 6.4. Another sequential process of building a reserve network, illustrated for Florida. The sequence shows managed areas existing today, which constitute 3,206,050 ha or 22% of the state; what might exist in about 10 years if current land protection initiatives are successful, totaling 5,692,923 ha or 39% of the state; and what might exist in 100 years if more ambitious protection initiatives are successful, totaling 8,658,244 ha or 59% of the state. From Noss (1996b) based on maps and data from Noss (1987c), Cox et al. (1994), and The Nature Conservancy/Florida Natural Areas Inventory (unpublished).

authors attributed this biotic impoverishment to housing develop-ment, although lack of ecological management might have also been a factor. A well-considered conservation plan will account for the nature and impacts of adjacent development on reserves.

Direct human impact from visitation is another potential impact on the biota in reserves. Studies have shown significant changes in ground-layer vegetation in the vicinity of hiking trails (e.g., Adkison and Jackson 1996). A woodland park in Boston lost 37 percent of its native plant species from 1894 to 1993 and gained many exotics; in this case the cause seems to be an increase in human use of the park, accompanied by more roads and trails, thinning of the forest, and trampling of sensitive areas (Drayton and Primack 1996). Other research is showing depression of bird activity up to a couple of hundred meters from trails and abandonment of areas by wolves as human recreational use increases (M. Soulé and P. Paquet, pers. comm.). Thus, if not carefully managed and controlled, human activities within or adjacent to reserves can severely limit their ability to maintain target species and other elements of biodiversity.

We recognize the key role that human use and recreation play in public support for conservation plans and as a political incentive to develop these plans. In many cases the entire public justification for a conservation plan may be the open space and recreational value of the set-aside land. Nevertheless, the necessity of managing and mitigating human impacts—especially those from active recreation—to ensure that the biological goals of the reserve design are attained must be made clear to all participants in the planning process from the beginning, so that there are no false expectations. Again, we emphasize that unless biological values have the highest priority in designing and implementing plans, conservation goals are likely to be undermined.

Sustenance of Ecological and Evolutionary Processes

One of the more prominent trends in the ecological sciences in recent years is an increased attention to processes. The ecological literature is burgeoning with papers on disturbance regimes (see Pickett and White 1985), nutrient cycling, hydrology, dispersal, and biotic interactions, while the literature in evolution has been more and more concerned with molecular genetic processes and the mechanisms through which populations differentiate over time. On the other hand, explicit consideration of ecological and evolutionary processes in conservation planning and land management has been rare. But there are exceptions, and the trend may be shifting toward greater awareness of the importance of processes.

For example, planners in the HCP for the Coachella Valley fringe-toed lizard were immediately confronted by the fundamental threat to the species: disruption of the aeolian transport of fine sand that maintains the dune habitat required by the lizard (see chapter 5). Hence, the process of sand deposition became a major focus of the HCP. In a case illustrative of the onerous choices often accompanying conservation planning, participants were forced to choose between two sand sources. One source was less disturbed than the other, but it involved dozens if not hundreds of landowners. It would have taken many years to secure. The less optimal sand source was possible to protect immediately. Monitoring and implementation of the plan, including refinement of the ecological model, has led scientists to the conclusion that the source previously protected is not sufficient to assure maintenance of the dune habitat. New sites are being considered for protection to enhance the opportunities for sand transport to the lizard preserves (Barrows 1996).

We urge increased attention both to biotic and abiotic processes in conservation planning. We caution, however, against an emphasis on processes at the expense of organisms. One danger in process-focused conservation planning is that processes may take on a life of their own, independent of the biotic elements of original concern. As Soulé (1996:59–60) has warned, "The processes of ecosystems are universal, but the species are not. . . . It is technically possible to maintain ecosystem processes, including a high level of economically beneficial productivity, by replacing the hundreds of native plants, invertebrates, and vertebrates with about 15 or 20 introduced, weedy species." For example, foresters are technically capable of replacing an old-growth forest with a tree farm that continues to photosynthesize, cycle nutrients, and protect the watershed, perhaps more efficiently in some respects than the original forest. But the biological value of the tree farm is much lower (Noss and Cooperrider 1994).

What planners must do is determine the kind of biological system they wish to perpetuate (with some flexibility, of course, because Nature is always changing) and then identify the key ecological processes that maintain the habitat structure and species composition of that desired community. The processes of concern may include fire, flooding, wind transport of sediments (as in the Coachella Valley example), herbivory, predation, pollination, seed dispersal, and many more. For example, many floodplain species require regular cycles of scouring by water and deposition of sediments in order to reproduce. Some pine species, including most races of lodgepole *(Pinus contorta)*, jack *(P. banksiana)*, and sand *(P. clausa)* pines, possess serotinous cones and require the intense heat of stand-replacing fires to open the cones and permit regeneration. Other pines, notably longleaf *(P. palustris)* and ponderosa *(P. ponderosa)*, are injured or killed by intense fires but require frequent, low- intensity ground fires to prevent competing species from dominating the site. In other cases, although the process of predation itself is entirely natural, exotic or often even native predators or parasites can eliminate populations of imperiled species, especially in fragmented and heavily disturbed landscapes. In such cases predators and parasites must be controlled (e.g., by trapping of cowbirds or house cats threatening endangered songbirds). Thus, maintaining the desired community is not simply a question of keeping processes operating; rather, it requires maintaining processes within the range of variability that the native species of the community have experienced during their evolution.

Maintaining ecological processes at appropriate levels usually requires active management and, in many cases, restoration. In most if not all conservation plans, we cannot count on natural processes operating effectively if we establish reserves and then leave them entirely alone. This problem arises largely because many natural processes operate on spatial scales much more vast than our reserve networks. For example, the hydrological system that once maintained the unique natural communities of the Everglades begins in a chain of lakes just south of Orlando and encompasses most of South Florida. Channelization, diversions, and other water projects have altered this hydrological system to the point where even the large reserves of Everglades National Park and Big Cypress National Preserve are significantly altered, and populations of wading birds and other species have plummeted over the past century. In the West, construction of the Glen Canyon Dam on the Colorado River has eliminated the floods that once swept through the Grand Canyon, downstream. The impact of this alteration has been felt through the gradual erosion of beaches and bars in the canyon, removing favored camping sites for visitors and habitat for native species.

Fire is another natural process that operates on large scales. Fires in some kinds of ecosystems, such as many western coniferous forests, affect hundreds of thousands of hectares or more. Within the heterogeneous fire mosaics that once were imprinted on these landscapes were many unburned refugia, from which species could recolonize burned areas. Now, stand-replacing fires in small reserves may leave no refugia, and populations may be extirpated by even "natural" fires. The Laguna Canyon fire in 1993 in southern California, which was apparently human-set, burned such a large area that it left only one small refugium of unburned cactus for the coastal cactus wren. It remains to be seen whether the wren can recolonize its regenerating habitat in the currently fragmented landscape. For ecosystems that require frequent fire, such as many grasslands, pine forests, and savannas, lightning simply does not strike small reserves often enough to burn them regularly. In all these cases natural processes must now be mimicked through such active management as simulated floods and prescribed fires. For these reasons a habitat-based conservation plan should contain specific provisions on how natural processes will be sustained through management. In many cases, especially in fragmented and degraded landscapes, the active restoration of processes in a reserve system—combined with control of exotic or other noxious species, harmful recreation, and other direct threats—is likely to provide the greatest incremental positive change in habitat values.

Ecological Monitoring and Adaptive Management

As noted earlier, adaptive management uses information derived from research and monitoring to revise conservation plans as part of a continual, adaptive feedback loop (Holling 1978; Walters 1986; Noss and Cooperrider 1994) (fig. 6.5). Adaptive management is where science enters the *management* or *implementation* phase of conservation, as opposed to simply being involved in the *planning* phase. Again, we emphasize that planning and implementation should be tightly integrated and that science is the key to this integration.

Given the crisis atmosphere in which most conservation plans are developed, the implementation phase is probably where the increased involvement of scientists is most needed. In many cases scientists have provided input on definition of the planning region, the target species to be considered in the plan, biological survey and research guidelines, population viability concerns for various species, reserve design, and other topics prior to implementation of the plan. Once a plan has been implemented, scientists usually have been less involved, and little additional research or ecological monitoring has taken place. (We recognize exceptions to this generalization, such as the Coachella Valley HCP described earlier.) Or, if research or monitoring has been conducted, it has had a minor influence on management. This nonadaptive approach, which has been called "linear comprehensive management"

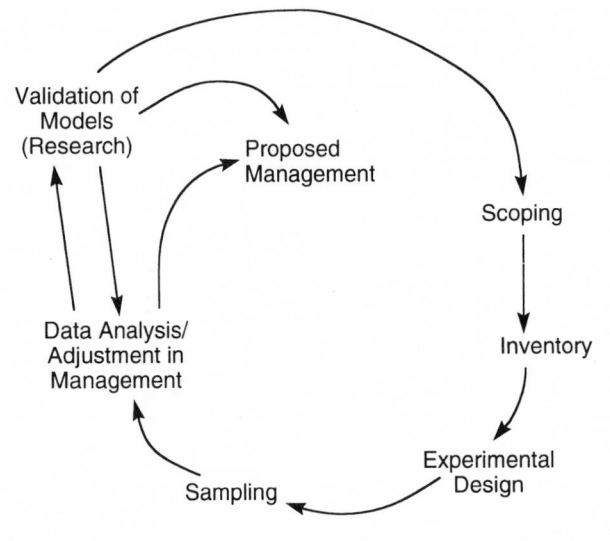

FIGURE 6.5. The adaptive management and monitoring cycle. From Noss and Cooperrider (1994).

(Bailey 1982; Noss and Cooperrider 1994), carries many naive assumptions about our level of understanding of ecosystems and includes the belief that human activities (including management actions) pose minor risks or have only short-term or reversible consequences. In contrast, adaptive management is experimental and cautious and leaves options for change open (table 6.1). It is based on monitoring, meaning the periodic measurement or observation of a process or object (Noss and Cooperrider 1994). Although the two types of management described in table 6.1 are "straw men" in the sense that they are two extreme ends of a spectrum, we have seen many examples that fit these extremes (especially the linear comprehensive model).

In the context of conservation planning, adaptive management requires a commitment of science to the conservation process in perpetuity. Rigorous scientific standards are applied both to research and monitoring, and research and monitoring are inextricably linked rather than treated as separate activities (Noss 1990; Noss and Cooperrider 1994). Information coming in from on-site monitoring (e.g., population trends of target species, responses of a community to prescribed burning) is combined with data from research (e.g., population genetics or dispersal behavior of target species) to inform and revise site-specific and regional management plans. This synthesis of

TABLE 6.1.

Comparison of adaptive management and linear comprehensive management in the context of habitat-based conservation planning (from Noss and Cooperrider 1994). We recognize that the attributes given as examples represent extreme cases of each type of management.

	LINEAR COMPREHENSIVE MANAGEMENT	ADAPTIVE MANAGEMENT
Concern with ecosystem	Minimal concern with ecosystem due either to belief in human ability to manipulate or restore ecosystem or to lack of concern with ecosystem degradation	Recognition of overriding value of ecosystem and necessity of conserving a properly functioning ecosystem for many reasons
Knowledge of ecosystem	Assumes that ecosystems, ecosystem processes, and effects of humans on them can be easily understood and predicted by traditional reductionist science	Recognizes that ecosystems and ecosystem processes are beyond human ability to understand except in the most rudimentary way and that effects of human actions on them are to a large extent unpredictable
Method of predicting effects of human actions	Emphasizes traditional reductionist science aided by modern high-tech tools, such as computer models	Emphasizes using experience to learn incrementally, starting with small-scale experiments and slowly and cautiously gathering new knowledge

Risk	Assumes that human actions pose little threat to ecosystem or that such risks are not a concern	Emphasizes minimizing risk to ecosystem
Scale—Spatial	Assumes that knowledge about ecosystems and effects of humans on them can be extrapolated across large regions; bases management on assumptions that effects are local	Recognizes that local ecosystems are unique and that extrapolating across large regions is risky; recognizes that all ecosystems are connected and that local actions can have major effects on other or larger regions up to the global
Scale—Temporal	Assumes that effects of human activities on ecosystems are generally short-term and reversible	Recognizes that effects of human activities may be long-term and/or have time lags before effects are observed
Learning/monitoring	Assumes that learning from management actions is not necessary; monitoring not necessary since outcomes are predictable	Recognizes that careful and systematic monitoring is essential in order to learn how to manage ecosystems sustainably

research and monitoring will not be easy to achieve and should not be treated casually.

Ongoing research and monitoring will be financially costly—though they need not be prohibitively so—and sufficient funds for these activities must be authorized and set aside as part of a management endowment before a plan is considered for approval. Otherwise, there is little opportunity to learn from experience in any rigorous way and the conservation plan has an elevated probability of long-term failure. We see generous funding for adaptive management as the most useful way to prevent negative impacts of "no surprises" policies. Another way of learning from experience would be to require periodic "audits" of selected or random plans by the Department of Interior or a designated entity, in order to determine how well they are achieving their goals (M. Bean, pers. comm.). Because of the idiosyncracies of each ecosystem, however, we do not have much faith that the success or failure of one plan will tell us everything we need to know about the status of a similarly structured plan in another region. Hence, we consider audits a worthwhile addition to, but not a substitute for, rigorous monitoring and a true adaptive management capability in each plan.

We stress that it is much easier to revise management practices than to revise reserve design, once a conservation plan has been approved and is being implemented. Although there will be cases where adaptive management points to a need for increased land protection, enlarging or otherwise changing reserve boundaries may require administrative or congressional approval (for public lands) or substantial funds for land acquisition (for private lands). In the latter case, when assurances to landowners have been made to secure their cooperation in conservation planning, any new costs must be at public expense. Exceptions would include voluntary land trades or donations, but these opportunities will probably be rare. In most cases we expect that little private land will be added to a reserve system after a plan is approved. Adaptive management on public lands and on private lands that were designated for conservation in a plan must consist mostly of changes in management practices and public-use (i.e., recreation) policies.

These considerations highlight the important role of public lands in planning regions composed of a mixture of public and private tracts. It is conventionally assumed that the public lands will function as reserves in a conservation plan; hence, less private land needs to be protected. But what if, at some later date, new timber sales, increased live-stock grazing or mining, or new areas for off-road vehicle use are proposed for those public lands? Barring compensatory increases in the amount or quality of private-land habitat, activities that would degrade public lands must be prohibited in the plan. In fact, in most cases habitat will need to be improved on public lands through ecological management if these areas are to function as true reserves. Public lands in the United States are generally not managed for biodiversity, and some areas are biological wastelands (Noss and Cooperrider 1994). Obligations of public land managers should be clearly stated in each memorandum of understanding (MOU) or other agreement with public agencies participating in a conservation plan.

The difficulty and cost of changing reserve design after approval of a conservation plan underscores the need for the initial plan to have the highest scientific defensibility possible, and to risk erring on the side of overprotection. It might be easier to dispense with some protected land later, if it turns out that more was protected than necessary to achieve conservation objectives, than to buy back and restore land already developed. In some public land situations, however, increasing protection may be easier. For example, it may be more difficult to develop land approved for "permanent" protection, for example in congressionally designated wilderness, than to restore a partially logged forest and add it to a reserve system. In either case prudence suggests that erring on the side of overprotection should be preferred to overdevelopment. Species lost locally can be reintroduced successfully only if suitable habitat conditions remain or can be restored; species lost globally can never (given existing technologies) be revived.

Ecological monitoring is a science of its own that we treat only minimally here (see Noss and Cooperrider 1994 and references therein). The commitment to monitoring, on the part of both public land managers and stewards of private land, has often been less than needed to manage land intelligently. Adaptive management plainly cannot occur without monitoring. Of fundamental importance is selecting the right attributes or "indicators" to monitor. Indicator selection for assessment and monitoring of biodiversity or ecological integrity has been extensively discussed in the recent literature (see Karr 1981, 1991; Noss 1990, 1995; Turner and Gardner 1991; Woodley 1993), and there is a plethora of indicators to choose from. Yet, the problem remains of selecting the best indicators for the situation at hand. One cannot monitor everything, but must start with a reasonably comprehensive list of possibilities (box 6.5). Criteria suggested to narrow the list of potential indicators to those that are most useful include (1) a validated relationship of the indicator to the phenomenon of interest; (2) convenience and cost-effectiveness of the indicator for repeated measurement; (3) ability of the indicator to provide an early warning of change or trouble ahead; and (4) ability of an indicator to distinguish changes caused by human activity from "natural" changes (see reviews by Noss 1990, 1995). A fifth criterion might be that the indicator must be straightforward enough that, if you must go to court to force a change in management, you could persuade twelve jurors that your data are compelling (S. Johnson, pers. comm.).

Although funding and staffing limitations in any conservation plan will restrict the number of indicators that can be measured, relying on one or a very few indicators is precarious. Unfortunately, indicator selection always will be hampered by our basic ignorance of ecosystems (Keddy et al. 1993). The best that managers can do is generate reasonable hypotheses about the controlling factors that maintain the communities and species of concern, based on available empirical data and theory, and select indicators with verified or highly probable relationships to those factors. The level of correspondence between the indicator and the property of interest (e.g., population viability of a target species) should be routinely validated by focused research.

To develop indicators at the population level, one must decide which of many rare, threatened, endangered, or otherwise sensitive species should be monitored (see chapter 5, box 5.1); for these target or indicator species one must then select the specific indicators or attributes of demography or genetic structure to measure (see box 6.5). The ability to detect population declines, specify the critical amount of habitat needed for persistence, or the level of habitat degradation leading to decline will always be limited. Simple census data are often inadequate, and the statistical power to detect a population decline decreases as the population becomes smaller (Taylor and Gerrodette 1993). Thus, the populations most in need of monitoring will be the most difficult to monitor in a way that provides information valuable for management. Furthermore, time lags in the response of demographic parameters to habitat degradation suggest that by the time a population decline is detected, even with excellent data derived from monitoring, it may be too late to take corrective action (Doak 1995). Therefore, detection of a statistically significant decline should not be a precondition for increased protection or management to aid a sensitive species—the burden of proof must be on those who would impose decreased protection (Taylor and Gerrodette 1993). A rigorous demonstration that recruitment exceeds mortality would provide that proof, temporarily at least, but habitat conditions known to affect the viability of the species (e.g., invasion by competing exotic species, or open-road density and use in the case of many terrestrial animals) must be closely monitored in all cases. Source habitats (where local recruitment exceeds mortality) must also be carefully distinguished from sinks (where local mortality exceeds recruitment).

The remarkable advance of techniques of remote sensing and GIS in recent years has made landscape-scale assessments and monitoring more attractive (see box 6.2). Mladenoff et al. (1995) found several landscape variables significant in predicting habitat suitability for wolves in the northern Great Lakes region; most important were road density and fractal dimension (a measure of habitat fragmentation). Nevertheless, many landscape metrics suggested for measurement by various authors have no validated relationship to biological phenomena of interest (Noss 1995) and may reflect fads. Schumaker (1996) tested several indices of landscape pattern that are often suggested as measures of connectivity. A valid measure of connectivity should correlate with dispersal success for particular species known to be threatened by habitat fragmentation. For the northern spotted owl, nine common indices of landscape pattern were poorly correlated with modeled dispersal success; however, a new synthetic index of pattern, called patch cohesion, predicted dispersal success well across a broad range of landscape types, territory sizes, and movement abilities (Schumaker 1996). Of course, we would not necessarily expect an index that works for one species to apply well to species with very different life histories. Thus, there is a tremendous need to test landscape indices against measured (not just modeled) demographic rates for a wide variety of species.

As noted by Christensen et al. (1996:681), "The design, development, and maintenance of monitoring programs requires commitment and long-term vision." Although we are enthusiastic about the concept of adaptive management and are pleased to see it become popular in conservation and natural resources management, we suggest that claims about the adaptability of conservation plans be scrutinized carefully. The ability of agencies, organizations, or other landowners to implement management that is truly adaptive requires three

Box 6.5.

Examples of measurable indicators of ecological integrity for potential use in monitoring programs and adaptive management (adapted from Noss 1995). Modification of these indicators to fit local conditions is essential, as is screening to select those indicators that have a validated relationship to the phenomenon of interest (e.g., viability of a target species or natural community) and can be measured repeatedly in a cost-effective way.

Landscape-Regional Level

Structural measures of patch characteristics

- patch size frequency distribution for each seral stage and community type, and across all stages and types
- size frequency distribution of late-successional forest patches (minus defined edge zone, e.g., 100–200 m)
- total amount of late-successional forest interior habitat in all patches and per patch
- total amount of patch perimeter and edge zone (also patch perimeter:area ratios, edge zone:interior zone ratios)
- patch shape indices (e.g., deviation from roundness)

Structural measures of patch dispersion

- patch density
- fragmentation (e.g., fractal dimension) and connectivity indices
- interpatch distance (mean, median, range) for various patch types
- juxtaposition measures (percentage of area within a defined distance from patch occupied by different habitat types, length of patch border adjacent to different habitat types)
- structural contrast (magnitude of difference between adjacent habitats, measured for various structural attributes)

Access, flow, and disturbance indicators

- frequency, return interval, or rotation period of fires and other natural and anthropogenic disturbances
- road density (km/km^2) for different classes of road and all road classes combined
- percentage of zone in roadless area (for different size thresholds, e.g., 1,000 ha and above, 5,000 ha and above)

Box 6.5. 195

- kilometers of roads constructed, reconstructed, and closed (seasonally and permanently) each decade
- amount of roadless area restored through permanent road closures and revegetation each decade
- density of airstrips, boat landings, and other access points, how frequently they are used, and how many people go in and out per landing
- density of livestock (or in some cases, historical density)
- energy, nutrient, water, and organism (including human) fluxes between and among habitats or zones

Community-Ecosystem Level

Structural measures

- frequency distribution of seral stages (age classes) for each community type and across all types
- average and range of tree ages within defined seral stages of forest
- ratio of area of natural habitat to anthropogenic or human-disturbed habitat
- abundance and density of key structural features (e.g., snags and downed logs in various size and decay classes) in either terrestrial or aquatic habitats
- spatial dispersion of structural elements and patches
- physiognomy, including foliage density and layering (profiles), and horizontal diversity of foliage profiles in stand
- canopy density and size and dispersion of canopy openings
- woody stem density in various size (dbh) classes
- diversity of tree ages or sizes in stand
- cover of native graminoids in open forest, grassland, shrubland, and tundra communities

Compositional measures

- identity, relative abundance, frequency, richness, and evenness of species and guilds (in various habitats)
- ratio of exotic species to native species in community (species richness, cover, and biomass)

Functional measures

- frequency, return interval, or rotation period of fires and other natural and anthropogenic disturbances
- areal extent of each disturbance event
- intensity or severity of disturbance events

- seasonality or periodicity of disturbances
- predictability or variability of disturbances
- invasion rates of weedy or opportunistic species (e.g., shrubs or purple loosestrife in wetlands, shrubs in prairies, exotic plants or birds in range-lands, exotic fish in streams or lakes)
- human intrusion rates and intensities
- nutrient cycling rates (for key limiting nutrients)

Composite indices
- Karr's index of biotic integrity (IBI), composed of twelve attributes of fish communities in three major classes (species richness and composition, trophic composition, and fish abundance and condition), or adaptations to other taxa or habitats

Species Level

Measures of genetic integrity (or lack thereof)
- heterozygosity
- allelic diversity
- presence/absence of rare alleles
- phenotypic polymorphism
- symptoms of inbreeding depression or genetic drift (reduced survivorship or fertility, abnormal sperm, reduced resistance to disease, morphological abnormalities or asymmetries)
- inbreeding/outbreeding rate
- rate of genetic interchange between populations (measured by rate of dispersal and subsequent reproduction of migrants)

Measures of demographic integrity (or lack thereof)
- abundance, density, cover or importance value
- fertility or recruitment rate
- survivorship or mortality rate
- sex ratio and age distribution
- health parameters (fecundity, individual growth rate, body mass, stress hormone levels, etc.)
- population growth and fluctuation trends
- distribution and dispersion of subpopulations or individual home ranges across the region
- trends in habitat components (varies by species)
- trends in threats to species (depends on life history and sensitivity of species; see fragmentation and other landscape indices above)

things: (1) adequate monitoring and integration of information derived from monitoring; (2) expeditious incorporation of information obtained from monitoring into management programs; and (3) managers and scientists cooperating with each other and the public to assure that expectations and values are incorporated into the process (Mattson et al. 1996). All three of these conditions are contingent upon agencies or other landowners having adaptive, open-learning mechanisms and sufficient resources, authority, and incentives to obtain and use information (Brewer and Clark 1994; Mattson et al. 1996). Whether agencies and landowners are capable of making the revolutionary switch from traditional "command and control" behavior to truly adaptive behavior that places highest value on sustainability of the ecosystem remains to be seen (Christensen et al. 1996; Holling and Meffe 1996).

Unifying Disparate Planning Efforts

Perhaps the greatest challenge—and need—in habitat-based conservation planning is, in each region, to link the variety of different agency and private initiatives into a unified approach to protecting imperiled species and ecosystems and reconciling economic activity with that protection. At any given time there are federally mandated conservation plans under Section 10 of ESA, state mechanisms such as NCCP and state-level endangered species and nongame programs, local government planning and zoning processes, and bioregional conservation strategies initiated by citizens. Some states, such as Florida, have a regional (multicounty) planning structure. In some cases coordination is needed with foreign governments or organizations for conservation problems that cross national borders (e.g., see the special section on conserving large carnivores in the Rocky Mountains of the United States and Canada in the August 1996 issue of *Conservation Biology*). If all these processes involving many agencies and groups at different scales are not linked in some manner, they will at best be inefficient and waste resources. At worst, they will conflict with each and result in poor decisions about both conservation and human infrastructure, leading to further costs, losses of biodiversity, and degradation of the quality of human life. We will not offer a detailed plan for a unified approach in this book, but we consider it valuable to highlight the issue and offer some thoughts on its resolution.

The political obstacles to local, regional, state, and federal cooperation in land-use and conservation planning are daunting. To begin to
address the issue, we suggest that the common ground among all
these approaches rests in conservation biology. Jurisdictions, be they
local, state, or federal, that are involved in conservation planning are
defined by political and social boundaries, boundaries that Nature and
natural science do not recognize. A comprehensive and scientific approach to conservation planning not only identifies the opportunities
for cooperation, it literally forces jurisdictions to consider the effects
of each other's efforts in order to avoid conflicts. For example, as noted
earlier, one of the Balcones Canyonlands Conservation Plan's target
species is the golden-cheeked warbler. Conserving this species requires consideration not only of its nesting habitat in the Texas Hill
Country, but also its overwintering habitat in southern Mexico and
Central America. A conservation plan in Texas that satisfied all this
bird's requirements for survival there would be of little long-term
value if no attempt were made to protect the species in its wintering
habitat (and, to date, there has been little or no attempt). This kind of
problem is common to all migratory and nomadic species. As another
example, the white ibis *(Eudocimus albus)* is not migratory, but populations have undergone at least four major geographic shifts in the
southern United States between 1930 and 1993, presumably in response to weather-induced changes in habitat conditions (Frederick et
al. 1996). Conventional conservation planning, even at the scale of regions as commonly defined, will not suffice for species such as these.
Rather, coordination of monitoring and conservation efforts at very
broad, often international, scales is needed.

Coordination should start in science with a comprehensive within-
regional and among-regional analysis, along the lines suggested in
this book and attempted for ecosystems such as old growth in the Pacific Northwest (Thomas et al. 1990; Murphy and Noon 1992) and terrestrial ecoregions of Latin America and the Caribbean (Dinerstein et
al. 1995) and the states and ecoregions of North America (Noss and
Peters 1995; Ricketts et al. 1997). Regions of continental or global significance, in terms of being biologically distinct (e.g., high endemism)
and at high risk of impoverishment, may warrant stronger conservation standards than regions less distinct or threatened (see box 6.6).
This does not mean that standards for scientific scrutiny should be
lower in lower-risk regions, only that regions that have more to
lose—and are at greater risk of losing it soon—must be treated with a

commensurately high level of caution and will usually require proportionately more protected area.

The interactions among the participants in a conservation plan need much better coordination than has usually occurred in the past. Rather than pursuing a piecemeal approach, such as project-by-project permitting, regulatory jurisdictions should use an exercise in regional conservation biology as the foundation for cooperation, with each assuming responsibility for the appropriate spatial scale of their involvement. City and county planning agencies and park districts must not only consider their own obligations; they must understand and react to how their local area fits into the context of the broader regional plan, which is the scale at which state and federal agencies will usually fulfill their major responsibilities. Further, such plans should be tightly coordinated with recovery plans for listed species, which are again the responsibility of the federal government (FWS).

We suggest that a well-coordinated approach—although challenging to develop—not only will be more effective for conservation in the long run, but it will also dramatically increase flexibility for conservation alternatives and perhaps even reduce costs because of increased opportunities for sharing of responsibilities. For example, there may be little need for a local jurisdiction to go to great lengths and expense to protect a particular habitat patch if a larger, more biologically valuable patch of the same type is in a reserve in the adjoining county. Five remnant pairs of a particular bird species, while appearing locally rare, may not require protection if there are 300 pairs nesting in the next county. On the other hand, if the county discovers that the best remaining example of a particular natural community or the largest remaining population of a rare species is within its boundaries—a discovery only possible when the status of each species and community is considered across its range—then the county has an added responsibility to protect it (perhaps with contributions from its neighbors). Similarly, the need to provide landscape linkages to other counties and regions, in order to provide for adequate dispersal of wide-ranging species, is only evident on a regional or interregional scale of analysis. These examples highlight the need to consider *context* as well as *content* in conservation decisions (Noss and Harris 1986) and to consider both at multiple scales.

At some level regional cooperation must be administered and coordinated. Depending on the boundaries of the planning region, either the state or the federal government (if two or more states are

Box 6.6.
Selecting Priority Ecoregions

A couple of decades ago biologist Jared Diamond wrote that "conservation strategy should not treat all species as equal but must focus on species and habitats threatened by human activities" (Diamond 1976). Although a highly restricted focus on only the most imperiled species and habitats has been properly criticized because it is not proactive, scientists agree that prioritization in conservation is essential. Furthermore, the logic underlying attention to the most imperiled elements is unassailable: if we don't protect them now, they will soon be gone. We will always need an Endangered Species Act and the fine filter of protecting rare species. We also need greatly increased attention to the coarse filter of protecting endangered ecosystems—those that have declined the most, are rarest, have the most imperiled species associated with them, and/or are at greatest risk of further losses (see chapter 1, box 1.1).

We can extend the coarse filter strategy upward to still larger ecosystems, such as ecoregions. Large regions that are biologically distinct (e.g., hot spots of endemism), have many endangered species and ecosystems (e.g., natural communities) within them, and are at high risk of losing biodiversity quickly should be of higher priority from a continental or global perspective than ecoregions that are less distinct or threatened. All regions and their native flora and fauna are important and deserve concern from conservationists, but those regions that have the most to lose and will lose it soon unless corrective action is taken need urgent attention. From the standpoint of habitat-based conservation planning, standards for approval of plans in high-risk ecoregions should arguably be higher than in low-risk ecoregions. High-risk ecoregions will usually require more protected area in order to maintain their biodiversity.

Two studies, one by Defenders of Wildlife and one by World Wildlife Fund (WWF), have ranked large regions in North America in terms of their conservation value and risk. The Defenders of Wildlife study (Noss and Peters 1995) restricted its attention to the United States, and because many decisions and programs of great importance to conservation take place at a state level, it analyzed threats to ecosystems and their associated species by state boundaries. An overall risk index (see fig. A) was calculated by combining three factors: ecosystems at risk, species at risk, and risk from development. The ecosystem risk index was based simply on the number of the most highly endangered ecosystems nationally (table 1.2) that occur in each state. The species risk index is a

BOX 6.6. SELECTING PRIORITY ECOREGIONS   201

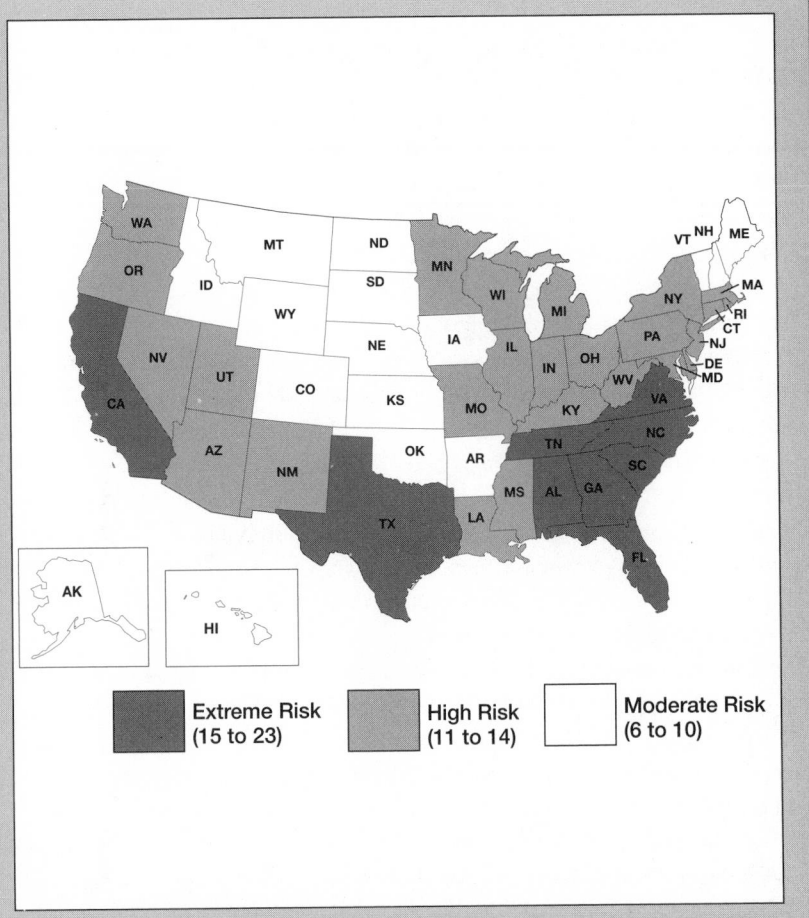

Figure A. The overall risk of states in the United States to biotic impoverishment. An overall risk index
was calculated for each state by combining three factors: number of highly endangered ecosystems;
percentage of species (vertebrates, aquatic mussels and crayfish, and vascular plants) in the state
considered globally imperiled (G2) or critically imperiled (G1) by The Nature Conservancy; and development risk, based on human population density, percentage of state in urban and agricultural
use, rural road density, and change in these factors during the decade 1982–1992. Adapted from
Noss and Peters (1995).

measure of the percentage of a state's species that are imperiled—specifically,
the proportion of vascular plants, vertebrates, and aquatic invertebrates (mussels and crayfish) that are ranked as G1 (critically imperiled globally) or G2 (imperiled globally) by The Nature Conservancy. The development risk index is
composed of two subindices: development status and development trend. A

state's development status was assessed by four equally weighted measures: human population density, percentage of state developed as of 1992, percentage of state in agricultural land, and rural road density. The development trend subindex measured the rate of development in recent years and was determined by the four factors: the number of people added to each square mile during the decade 1982–1992, the percent change in population density during this period, the percentage of the state that became developed during the decade, and the percent change in the total amount of developed land during the decade (Noss and Peters 1995).

The results of the Defenders of Wildlife study are evident from the map of overall risk. The southeastern states (Florida was highest of all states), Texas, and California and Hawai'i (tied for second highest risk) have the highest number of imperiled species and ecosystems and are losing biodiversity to development most rapidly. The urgency of conservation planning in these states is demonstrably high. Similarly, a study using county-level distributional data for endangered species to determine hotspots of threatened biodiversity found the highest concentrations in Hawai'i, southern California, the southeastern coastal states, and southern Appalachia (Dobson et al. 1997).

The WWF study (Ricketts et al. 1997) took the more ecologically realistic approach of analyzing risk by ecoregional, rather than political, boundaries and considered all of the United States and Canada. Ecoregions are relatively large areas of land or water that harbor characteristic species, communities, ecological phenomena and processes, and environmental conditions. Patterns of biodiversity are better reflected in ecoregional than political boundaries. The goals of the WWF assessment were to (1) identify ecoregions that support globally outstanding biodiversity and emphasize the global responsibility to protect or restore them; (2) assess the types and immediacy of threats to North American ecoregions; (3) identify appropriate conservation activities for each ecoregion based on its particular biological and ecological characteristics, conservation status, and threats; (4) encourage decision makers, conservation planners, and the public to adopt an ecoregion-based approach to conservation; and (5) provide a broad-scale framework so that conservation agencies and groups can position their activities within a continental and global context, resulting in more effective allocation of limited conservation resources.

The 116 terrestrial ecoregions identified for the United States and Canada (ecoregions for Mexico are currently being analyzed) were divided into ten Major Habitat Types to ensure good representation of terrestrial ecosystems, to compare only similar ecological systems, and to tailor analytical criteria to the

Box 6.6. Selecting Priority Ecoregions   203

particular patterns and sensitivities characteristic of different Major Habitat Types. Regional and taxonomic experts assessed the biological distinctiveness and conservation status of each ecoregion at a workshop in August 1996. Biological distinctiveness was determined through an analysis of species richness, endemism, distinctiveness of higher taxa, unusual ecological or evolutionary phenomena, or global rarity of Major Habitat Types. Conservation status was based on an assessment of such landscape and ecosystem-level features as habitat loss, habitat fragmentation, the size and number of large blocks of habitat, the degree of protection, and current and potential threats. Different combinations of biological distinctiveness and conservation status were used to prioritize ecoregions for conservation action and identify the most appropriate suite of conservation activities to be undertaken within them (fig. B). A summary map illustrates ecoregions of highest conservation concern (fig. C).

Some ecoregions were recognized as globally outstanding because of extraordinary endemism and richness such as the southeastern conifer forests, the Klamath-Siskiyou forests, the Appalachian forests, and the Chihuahuan Desert, while others were highlighted because of such ecological phenomena as the

MATRIX FOR RECOMMENDED CONSERVATION ACTIONS BY ECOREGION

	EXTINCT	CRITICAL	ENDANGERED	VULNERABLE	RELATIVELY STABLE	RELATIVELY INTACT
GLOBALLY OUTSTANDING		I	I	I	III	III
REGIONALLY OUTSTANDING		II	II	II	III	III
BIOREGIONALLY OUTSTANDING		IV	IV	V	V	V
NATIONALLY IMPORTANT		IV	IV	V	V	V

I. Globally outstanding ecoregions requiring immediate protection or restoration.

II. Regionally outstanding ecoregions requiring immediate protection or restoration.

III. Rare opportunities for conservation of globally or regionally outstanding ecoregions.

IV. Nationally important ecoregions requiring protection or restoration.

V. Nationally important ecoregions requiring proper management for biodiversity conservation.

Figure B. Matrix used by World Wildlife Fund to assign North American ecoregions to categories of conservation priority, based on each ecoregion's biological distinctiveness and conservation status (see text). No ecoregion is "written off" for conservation attention in this approach, but some ecoregions are recognized as demanding more urgent and stronger protective and restorative actions than others. From World Wildlife Fund.

presence of intact migrations of caribou or intact predator assemblages. In addition to other biodiversity features, some ecoregions were assessed as globally outstanding because of the global rarity of their habitat type, including the temperate rainforests of the Pacific Northwest and the California coastal sage. Temperate grasslands have been largely eliminated, while many forested ecoregions are highly threatened. The only relatively intact ecoregions occur in the boreal and arctic regions.

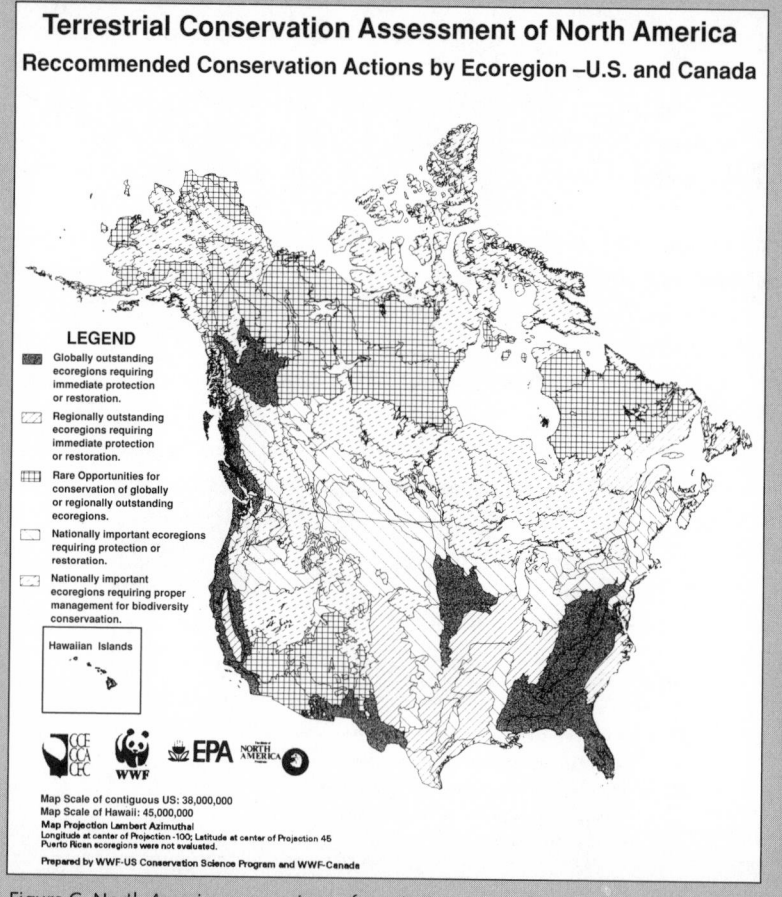

Figure C. North American ecoregions of greatest conservation concern, based on considerations of biological distinctiveness and conservation status. From World Wildlife Fund.

involved) is the most appropriate entity to perform this function. If the planning region crosses international boundaries, intergovernmental committees should take the lead. One cannot assume that the federal government is more likely than a state to have sufficient financial resources and a broader perspective to administer a conservation plan. In some cases the state may have more resources (e.g., the land acquisition program of the state of Florida dwarfs the land acquisition budget of the FWS and National Park Service for the entire nation). A lack of resources within the FWS may be the major cause of the chronic complaints about the Service's inability to respond in a timely manner to requests for information, advice, and assistance (M. Bean, pers. comm.).

We do not suggest that the state or federal government should be the controlling stakeholder in the planning process or be responsible for local planning decisions. This has rarely worked in the past and in some cases has sparked local resentment. It may be preferable for a state to delegate tasks and authority for the planning and scientific process to local jurisdictions that have the infrastructure and local clout to accomplish them, while performing a coordinating role that ensures that local entities have sufficient information and a broad perspective. In cases where local entities fail to consider their plans within a broad context, apply the best available science, or sincerely seek to attain conservation goals, the state or federal government can step in and exercise regulatory authority. This strategy has never been implemented fully on the scale we are proposing, nor with a foundation in science that will maximize local flexibility. However, the California NCCP program is in part a test of this idea, a large experiment where the FWS holds the regulatory authority with the listing of the California gnatcatcher and delegates conservation responsibility to the state of California, which in turn has given planning authority to city and county governments through a set of process guidelines. This experiment in intergovernmental, multiscale coordination has yet to be completed, and it is too early to say whether it will succeed or fail; but it has yielded some valuable lessons about comprehensive conservation planning (see previous chapters).

Regional coordination of disparate planning efforts is unlikely to occur merely through the good will of the participants. At some point there need to be both incentives to participate and formalization of the process through the regulatory system (i.e., disincentives for not participating). Incentives may be direct, such as priority for planning and land acquisition money given to participants in a regional conservation strategy and assurances of regulatory certainty upon completion (see box 2.1). Or incentives may be indirect, such as the economy of scale and biological flexibility created by conservation at a regional scale. More effort should be placed on identifying incentives and both enabling and promoting them.

One of the key features of a truly unified approach to conservation will be a reconciliation of species-level and ecosystem-level strategies and management plans (see chapter 1, Box 1.2). Because biodiversity occurs at several levels of organization, a comprehensive conservation strategy should also concentrate on several levels, including genes, populations, species, communities, ecosystems, and landscapes (Noss and Harris 1986). Trying to address all these levels at once would be confusing and could result in chaotic, fragmentary management. Like many authors, we recommend the ecosystem—especially at a landscape (kilometers-wide) scale—as the appropriate central focus, around which programs for elements at other levels can be organized. As part of the ecosystem approach, species with special needs, of great ecological importance, or that are highly vulnerable to human activities can still receive the individual attention they need to survive (box 5.1). Several habitat-based conservation plans under development (see the case studies in chapter 5) show potential for reconciling species and ecosystem conservation needs.

There will be times when the needs of particular species of concern appear to conflict with goals for maintaining the integrity of the ecosystem, for instance as expressed by native species composition. As a case in point, the endangered giant kangaroo rat *(Dipodomys ingens)* in the Carrizo Plains area of California has what has been characterized as a mutualistic relationship with exotic plants. Burrow digging and other activities of the kangaroo rat favor the establishment of exotic plants, which in turn produce a favored food (large seeds) of the rat (Schiffman 1994). A narrow conservation plan focused on the kangaroo rat could promote exotic plants, whereas a plan focused on the integrity of the natural community or ecosystem might call for eradication of the exotic plants, with possible harm to the rat. Although this case may seem irreconcilable, it is obvious that the rat at one time did well enough with the natural flora, and a total dependence of the rat on exotics is highly unlikely (S. Johnson, pers. comm.). Restoration of the community may be difficult, but it is not impossible. We do not feel that maximizing a population of a listed species is an appropriate goal if it comes at the expense of the integrity of the larger ecosystem of which it is a part. If one chose the rat over the natural community, many other native species could be affected negatively, continuing the downward spiral of biodiversity loss. On the other hand, to consign an imperiled species to extinction for the good of the community not only has parallels in fascism, but it could also result in an overall loss of biodiversity. In this case if the rat begins declining rapidly toward extinction, emergency measures to ensure its survival (e.g., supplemental feeding or captive breeding) might be necessary until the native flora and other aspects of the natural community can be restored. Conservation biologists and planners will often need to use their informed judgment to resolve such dilemmas.

There is much to learn about how to implement the principles and guidelines offered in this book, how they should be refined to apply to specific situations, and how they should be revised, replaced, or augmented with new principles and guidelines on the basis of experience. As suggested earlier, the optimal approach to improving conservation planning is to proceed with a series of case studies, which serve as demonstration projects or prototypes for future efforts. Prototyping has been shown to be a powerful means of innovation (Clark et al. 1995, 1996; Brunner and Clark 1996) and is a low-risk strategy for improving the process of conservation planning. Risks are low because each case study affects a relatively small geographic area and is therefore reasonably inexpensive and immune to political attack (compared, say, to implementing a national conservation strategy all at once). Case studies provide the only way for hypotheses derived from general principles to be tested in any rigorous fashion. Tangible results from specific projects are more likely to persuade people that the approach is useful than are mere proposals or plans, however logical or intuitively appealing (Clark et al. 1996). Through a collection of case studies—failures as well as successes—the conservation planning process can be improved, and the risks of conservation plans failing become ever lower. Hence, adaptive management can operate on a truly broad scale as we learn from experience.

7
CONCLUSIONS

Habitat-based planning would be an intelligent way to guide development and conservation even in a steady-state economy or in regions where wilderness still prevails and the pressures of human population and economic growth are slight. In these increasingly rare situations the opportunities for truly proactive conservation are most conspicuous—it is possible in such cases to do it right the first time. In other, longer-settled regions, such as the Corn Belt of the midwestern United States where virtually all of the landscape has been stripped of its native vegetation and converted to monoculture, the argument for planning of any kind seems less compelling. Here, conservation sensibility seems to be a century or two late. But even in these regions issues involving the last scraps of natural area could be addressed intelligently through the application of sound planning principles informed by science. Last scraps always deserve special care.

As it turned out, and not surprisingly in retrospect, habitat-based conservation planning arose not in the kinds of regions just described, but in those rapidly changing landscapes where the conflicts between humanity and Nature are most immediate and intense and where private interests have the most to gain or lose financially. In regions such as southern California and Florida, for example, we have the combination of two extremes: high species richness and endemism on the one hand and excessive human population and economic growth on the other. A substantial amount of natural habitat remains in some of these landscapes, but it is going fast, and many species have been listed or await listing under the ESA. In another kind of region, exemplified by the Pacific Northwest and Southeastern Coastal Plain, conservation planning for private forest lands has become urgent because heavy logging and conversion of natural forests to plantations on both private and public lands have placed many forest species—the best-known of which are the northern spotted owl and red-cockaded woodpecker—in danger of extinction. The environmental situation in both kinds of regions can be summarized by one word: crisis.

No one denies that habitat-based conservation planning was born in crisis. But on a more fundamental level, the need for habitat-based conservation planning arose because of our utter failure, both locally and nationally, to plan for sustainable economic activity and environmental protection at the same time. We have failed to realize that ecology and economics are derived from the same Greek root—*oikos,* or house—and have pursued these activities independently of each other. The lack of connection in the public mind between economic vitality and conservation is incongruous, considering that both ultimately contribute to human well-being. Poll after poll shows the enduring link between open space and perceived quality of life for human communities. Of course, open space may or may not be land of high biological value. Nevertheless, land trusts and other protection organizations are growing rapidly and are increasingly interested in promoting biodiversity with their purchases. This is happening at the same time that federal, state, and local governments are expressing interest in working together to create long-term, regional conservation plans. With these trends we are hopeful that a process born is crisis may help prevent future crises.

Regional, habitat-based conservation planning need not be a highly centralized, top-down endeavor. In much of California, for example, conservation planning has come full circle and has been returned to the local level as a partial substitute for a rigorous policy of growth management. Conservation action by local governments, which generally has been resisted by environmentalists, can in some cases work quite well. Press et al. (1996) noted that the scale of local and regional control matches the range sizes of many endemic species; land acquisition is the most promising approach for protecting these species; and at least some local governments and nongovernmental organizations possess the capacity to identify, acquire, and manage habitat for these species. Press et al. (1996) focused on four counties in central coastal California and concluded that local conservation initiatives provide an important complement to the ecosystem management programs of federal agencies and the regional and broader proposals of such private groups as The Nature Conservancy and The Wildlands Project. In Florida, counties are given the responsibility of developing comprehensive plans for growth management. These plans are required to contain a "conservation element." Although some counties have not taken their conservation planning responsibilities seriously, others have undertaken extensive studies of natural areas and rare species within their county boundaries (e.g., Alachua County, Duever et al. 1987) and are pursuing land conservation programs that will supplement the state's program, which is by far the largest land acquisition effort in the country.

The role of science and scientists in conservation planning—local or otherwise—logically should be central but remains problematic. Science is too often considered another in a list of competing interests in the planning process, is marginalized as a tool of the environmental community (a distinct danger when scientists are used by environmentalists as "hired guns"), or is used as a shield to hide partisan politics, philosophical objections to growth, or an ideological aversion to negotiating with industry or developers. Scientists may also unintentionally subvert the planning process by continually calling for more basic research on rare species or other topics in cases where enough is known to move forward and where money would be better spent on direct actions such as land acquisition or management. None of these views or roles of science contributes positively to conservation planning. Combined, they threaten to undermine science-based conservation.

If we are to create conservation plans that truly benefit imperiled species and ecosystems while promoting responsible economic activity, then science must become the foundation for decision making, not a special interest to be compromised or negotiated. This does not mean that all planning decisions must conform to the opinions of scientists—this is unlikely to happen and is not appropriate in a democracy. Nor does it suggest that planners wait for all the data before making decisions. Instead, it means that conservation decisions should be informed by the best scientific information, techniques, and analysis available, and that conservation management should be adapted through time as new information develops. Most important, scientists as well as planners must devote themselves first and foremost to the outcome of conservation plans—that is, effective conservation—not to the process itself (however many research opportunities it provides) and certainly not to their own special interests in obtaining funds or exercising power.

We have attempted to provide a reasonably thorough treatment of the role of science in conservation planning. Implicit in this discussion is our recognition that conservation planning, especially at the regional level, is a complicated mix of science, economics, social policy, and politics, and that science by itself will not provide final solutions for every problem. We believe, however, that objective integration of science into the decision-making process will produce stronger plans with less long-term controversy and a greater chance of providing landowners and other private parties with the predictability they desire. It is virtually certain that conservation planning will play an increasingly large role in the implementation of the ESA. Scientist have expressed their views on this subject to policymakers; these views, not coincidentally, correspond to many of the recommendations made in this book (box 7.1). Virtually everyone—not the least of which the species of concern—should benefit from scientifically informed and defensible plans. With emerging federal and state policies promising greater and greater assurances for private participants, it is all the more important that plans be created on a solid foundation of science. It is our hope that this volume will inform and improve the process of obtaining and applying that information, thus helping habitat-based conservation planning fulfill its promise.

We recognize that we have not answered definitively the question of how much science is enough to produce an acceptable plan. There is no ready answer to this question. We have submitted that an acceptable plan is a scientifically defensible plan. But the standards for scientific defensibility are not universally established. The number and eminence of scientists involved in the process, the necessary amount of data or sample sizes, the detail and sophistication of analyses, and the thoroughness of peer review needed to produce a defensible plan are not absolutes and will not be widely agreed on even among scientists. The scientific criteria for accepting or rejecting a plan cannot be set out in black and white—at least not in great detail—just as the criteria for accepting or rejecting a paper submitted to a scientific journal are not rigorously defined but depend on the informed judgment of peer reviewers and editors. Although a fair amount of subjectivity creeps into these processes, this is unavoidable and it is the way science as a human endeavor works.

With these caveats we offer a few comments on what constitutes a scientifically defensible and acceptable plan. An acceptable plan uses the available resources and data carefully to evaluate planning alternatives, including "null models" of no development and no conservation, in terms of their probable effects on the species and ecosystems of concern. The examination of alternatives should be conceptually and methodologically rigorous and should include map-based projections of the probable consequences of following each alternative. The preferred alternative would be the one that, given the available information, appears to have the highest probability of attaining all the stated conservation goals (population viability of target species, ecological integrity and native species composition of the community, sustenance of ecological processes, etc.) while also achieving social and economic objectives within these biological constraints.

But the process should not stop there. The preferred alternative—the plan—should be rigorously scrutinized and revised as new information becomes available during the planning process and as comments and other input from independent scientists and citizens are received. Once a plan is approved, revision and refinement continue through the processes of ecological monitoring and adaptive management, subject to the constraints of assurances made to private parties and of available funding. The plan should include an explicit framework (e.g., a "decision tree") for making specific decisions about land protection and management, including adjustments in management in response to new information. From a broader perspective, and in line with our endorsement of the case-study approach, each planning process should be well informed by the successes and failures of other plans. We suggest that a federal authority, such as the Department of Interior, be responsible for periodic audits and performance reviews of plans, so that this information can rapidly make its way to planners in each region. We further suggest that independent scientists be contracted by the federal government to assist in these assessments.

The proof of the value of conservation planning lies in the successful implementation of plans. The best designed, most scientifically credible plan in the world is worth little if not implemented as intended. We have noted that conservation planning is inherently a politicized process. To exclude politics not only would be difficult, it would be improper. Politics should mean that people of diverse interests are involved in the planning process; politics go astray when special interests on any side sabotage a process that has been agreed on by the majority of interests and shown to be generally valid. We have not addressed thoroughly in this book the problem of direct political interference in conservation planning (but see appendix 5B). This is a very real problem in some cases, and it will not be solved by scientists. We predict, however, that the more scientifically defensible a plan is and the more open it is to full public participation, the less easy it will be for politicians or others to subvert it.

In other cases it is not direct political interference that prevents

Box 7.1.

In February 1997 a group of scientists, led by Dennis Murphy, met at Stanford University to discuss habitat conservation planning and other initiatives for implementing the Endangered Species Act on private lands. The scientists focused on several key provisions of recent bills to reauthorize the Act. Their discussions resulted in a letter to the administration and Congress expressing a consensus viewpoint on these issues. The letter, with minor editorial corrections, is reprinted below:

A Statement on Proposed Private Lands Initiatives and Reauthorization of the Endangered Species Act from the Meeting of Scientists at Stanford University

When the Endangered Species Act was authorized in 1973, Congress charged the Departments of the Interior and Commerce to conserve the ecosystems upon which threatened and endangered species depend, and to do so "using the best available scientific and commercial data." Despite remarkable growth in our scientific understanding of the conservation needs of threatened and endangered species during the past two decades, controversy continues to surround the Act, especially as it affects the use of private land. The Act's provisions for the treatment of imperiled species on private land are of major conservation concern both because, according to some estimates, more than half of all listed species occur wholly on private land, and because listed species on private land are faring worse in general than those on federal lands.

Various bills recently introduced in Congress propose changes in the Act's provisions for treating listed species on private land. The private lands provisions proposed in draft legislation would modify the habitat conservation planning (HCP) language of Section 10(a) of the Act. The HCP process was designed to mitigate substantially the impacts of otherwise legal activities on listed species. However, many recent HCPs have been developed without adequate scientific guidance and there is growing criticism from the scientific community that HCPs have the potential to become habitat giveaways that contribute to, rather than alleviate, threats to listed species and their habitats.

The proposed new provisions have the potential to either improve or worsen the conditions of listed species on private lands, depending on whether or not habitat conservation planning and management are based on objective scientific evidence and methods. To provide guidance on the scientific implications of proposed private lands provisions, a group of nationally respected conservation biologists met at Stanford University in February. Among the undersigned are

BOX 7.1.   215

ecologists and geneticists with extensive experience in conservation planning for imperiled species. Our group includes individuals with widely differing positions on how best to achieve the goals of the Endangered Species Act. The diverse composition of our group should give weight to our conclusions.

In considering private land conservation planning initiatives, we restricted ourselves to five agenda items that recur in draft bills and ongoing discussions in congressional and conservation circles: (1) the "no surprises" policy, (2) multiple species conservation planning, (3) "safe harbor" initiatives, (4) prelisting agreements, and (5) small-parcel landowner initiatives. We understand that this is not an exhaustive list of potential private lands policies and programs. We also recognize that there is overlap among many of the proposed provisions; for example, the no surprises policy is often viewed as an obligatory component of the other proposed provisions.

As the following discussion makes clear, we believe that the current proposed private lands amendments to the Endangered Species Act will not further the Act's goals unless those measures are implemented in a scientifically sound manner. However, our group believes that with essential stipulations, "landowner-friendly" initiatives can assist in meeting our nation's goal of protecting its unique and valuable natural heritage.

No surprises

More aptly labeled "fair assurances" to landowners, no surprises policy promises that if private landowners protect targeted species under a Habitat Conservation Plan or the equivalent, they then will not have to underwrite future conservation requirements that may develop due to new information or changed circumstances. Should the species require further conservation efforts, the costs would be largely borne by the public rather than the landowners.

A no surprises policy is troubling to scientists because it runs counter to the natural world, which is full of surprises. Nature frequently produces surprises, such as new diseases, droughts, storms, floods, and fire. The inherent dynamic complexity of natural biological systems precludes accurate, specific prediction in most situations; and human activities greatly add to and compound this complexity. Surprises will occur in the future; it is only the nature and timing of surprises that are unpredictable. Furthermore, scientific research produces surprises in the form of new information regarding species, habitats, and natural processes. Habitat Conservation Plans, therefore, are inevitably developed and authorized under conditions of substantial uncertainty and may ultimately prove inadequate. Unless conservation plans can be amended, habitats and species certainly will be lost.

We appreciate that no surprises policy is not a guarantee that conservation

plans will not change, but a contractual commitment to shift some of the financial burden of future changes in agreements to the public. In that light, the following features should constitute minimum standards for HCPs with no surprises assurances. First, it must be possible to amend HCPs based on new
information, and it should not require "extraordinary circumstances" to do so.
Second, to underwrite program changes when parties other than the landowner
request and justify them, there must be a source of adequate, assured funding
that is not subject to the vagaries of the normal appropriation processes. We expect that the costs of fixing inadequate HCPs may be substantial. Third, mechanisms to ensure that long-term conservation plans will be monitored adequately
are essential. Monitoring habitat changes or ecosystem functions cannot substitute for the monitoring of target species. Moreover, new scientific information
from monitoring should be incorporated into management as that information
becomes available. Fourth, HCPs must clearly articulate measurable biological
goals and demonstrate how those goals will be attained under the plans. Plans
should not undermine the recovery of listed or vulnerable species. Fifth, assurances to landowners should only be extended for those targeted species for
which the plan articulates species-specific goals that further conservation in a regional context, rather than in a local, piecemeal fashion.

Multiple-species HCPs

Although Habitat Conservation Plans originally focused on individual species in
local areas, today many planners are finding it preferable (biologically and often
economically) to plan for multiple species over entire regions. In the absence of
scientifically credible recovery plans, multiple-species HCPs should clearly articulate conservation goals and must demonstrate their contribution to the conservation or recovery of targeted species. In addition, multiple-species HCPs should
assume an extra burden of rigor, requiring independent scientific review of
goals, design, management, and monitoring. There should be a standing body
of independent scientists to establish minimum scientific and management standards for multiple-species HCPs. The comprehensiveness of independent scientific review should be appropriate to the size and duration of the plan.

Multiple-species Habitat Conservation Plans cannot be based solely on the
distribution and extent of different habitat types because this information does
not yield effective predictions of the distribution and abundance of individual
species. Such HCPs, therefore, must focus on specific target species, such as endemic, listed, indicator, and keystone species. If one species is chosen as an
indicator of the status of another species of conservation concern, the plan
should validate the connection between them. Species that are critical for ecosystem integrity, whether or not they are listed as endangered or threatened,

Box 7.1. 217

should be among the indicators chosen. In addition, the viability of all target species "covered" by a plan must be considered in a greater regional context, often well beyond the boundaries of the planning area itself. Adequate distributional and ecological information should be made available to assess the plan's impacts on all covered species.

Multiple-species Habitat Conservation Plans must include adequate research and monitoring programs. The target species covered by the plan, such as endemic, listed, indicator, and keystone species, must be monitored individually. Plans also must include an adaptive management program, so that management can be improved in the light of new information obtained by monitoring or other means. As is the case for "no surprises," besides being amendable, multiple-species HCPs must have an assured source of funds to support potential amendments.

Safe harbor initiatives

Safe harbor initiatives encourage private landowners to increase the amount of habitat available to endangered species. In the past, many landowners have been reluctant to restore or enhance habitat for fear of incurring added regulatory burdens that will curtail future use of their property. Under safe harbor policy, the landowner is obligated to maintain only the baseline utilization of the property by the species prior to habitat improvements, which means that the landowner will be free to undo those improvements at a later date.

Most of our group believes that deleterious consequences to protected species from safe harbor initiatives will be infrequent and that safe harbors could prove to be an important inducement to overcoming landowner unwillingness to take actions beneficial to imperiled species. Nonetheless, two concerns should be addressed in safe harbor agreements. First, the concepts of "baseline population" and "utilization" require a clear definition. Sources of scientific uncertainty should be addressed in defining the baseline status of species, just as for the no surprises policy. The determination of the safe harbor baseline depends on reliable survey techniques and scientific interpretation. Second, some species may be better candidates for safe harbor agreements than others as a result of their distribution, resource needs, and habitat area requirements. Species are distributed across diverse landscapes with habitat areas of varying quality. In addition, species vary widely in their ability to move from one area of habitat to a neighboring one. Thus, we believe that the value of safe harbor agreements must be evaluated on a species-by-species basis. In the absence of scientifically credible recovery plans, safe harbor agreements should document their potential contributions to the conservation or recovery of target species within an entire region rather than on a single piece of private property.

Prelisting agreements

Under a prelisting agreement, a landowner would take actions to benefit an un-listed rare or declining species before it is listed. This has the potential to benefit species conservation because a species is afforded no protection on private land under the Endangered Species Act until it is listed. Nevertheless, prelisting agree-ments must not become an easy substitute for necessary listings.

Prelisting agreements often will be negotiated in the face of significant levels of scientific uncertainty—we know little about many of our listed species, less yet about many unlisted species. Because prelisting agreements should benefit species, we recommend an enhanced level of attention and critical review of the biological circumstances under consideration in proposed prelisting agreements. The federal government will have to deal with an inevitable shortfall of infor-mation; that situation can be partially corrected by (1) developing the most complete database possible to inform the decision, (2) clearly articulating how the prelisting agreement will benefit the targeted species, and (3) applying the necessary concomitants of the "no surprises" policy. The latter should include an ability to amend agreements, the availability of funding to support amend-ments, adaptive management with effective program monitoring, sufficient consideration of the regional planning context, and independent scientific review.

Small-parcel landowner initiatives

Considering the cost, complexity, and time required to complete Habitat Con-servation Plans and implement them, the idea of expediting the permitting process for small landowners is attractive. But we note that in many areas with imperiled species, private landholdings consist almost entirely of small parcels. When both large and small parcels are interspersed, the small parcels may con-tain most of the key habitat. Either way, the cumulative impacts of many small projects on imperiled species may be substantial. In addition, the relative impacts of small landowner activities vary greatly depending upon which endangered or threatened species live on their land. The loss of but five acres of remnant habitat could doom to extinction more than a few listed species. We are con-cerned that expediting the permitting process could come at a significant cost to species persistence.

Our group believes that any policy that allows for expedited HCPs should also require that such agreements not compromise the viability of targeted species within the planning region, and should explicitly consider and limit cu-mulative deleterious effects from incremental habitat losses. If a recovery plan exists, expedited HCPs must be consistent with the plan. Otherwise, to ensure coordination of existing and future HCPs, a regional analysis of species status should be required before any expedited HCPs or exemptions are considered.

Box 7.1.   219

Independent scientific review

While Habitat Conservation Plans and other conservation agreements that we have discussed above may offer promise for improved species protection on private and other nonfederal lands, serious questions remain about their effectiveness for long-term species conservation and recovery. Because many recovery plans and HCPs lack scientific validity, because the private lands proposals discussed above remain largely untested, and because endangered species protection and recovery must be based on the best available science, we believe that independent scientific review must become an essential step in the implementation of the Endangered Species Act. Such review should be carried out by scientists with no economic or other vested interests in the agreement. It is critical to start the review process early in the project, including the design phase.

Conclusion

Finally, while not strictly a "science" issue, we strongly agree that implementation of the Endangered Species Act would be immensely improved if funding were increased and agency staff were better trained. We agree that better enforcement of the Act's prohibitions by the U.S. Fish and Wildlife Service and the National Marine Fisheries Service would benefit listed species. We also agree that the Act's goals are compromised by conflicting laws and regulations that encourage actions that directly and indirectly contribute to species endangerment. And we concur that a wide array of incentives and inducements for better Act compliance by private parties could serve to benefit species conservation greatly if implemented in a scientifically responsible manner.

We hope that these observations and our scientific recommendations above will help Congress to enact legislation that will make the Endangered Species Act more acceptable to private landowners while strengthening the protection of species and habitats on private lands.

Dennis Murphy, *Stanford University*

Peter Brussard, *University of Nevada, Reno*

Gary Meffe, *Savannah River Ecology Laboratory*

Barry Noon, *U.S. Forest Service*

Reed Noss, *Oregon State University*

James Quinn, *University of California, Davis*

Katherine Ralls, *Smithsonian Institution*

Michael Soulé, *University of California, Santa Cruz*

Richard Tracy, *University of Nevada, Reno*

plans from being implemented successfully; rather, it is failure to provide sufficient funding or staff to carry a plan through to completion. Assuring the successful implementation of plans and the good will of participants also goes well beyond the role of science. Ultimately, conservation planning will work only with the active participation and vigilance of an informed and democratic citizenry.

We close with some final recommendations for scientists, citizen activists, developers or other industry, and agencies involved in conservation planning:

For Scientists

1. Do your best to set aside preconceived notions of what constitutes scientifically defensible conservation planning. There usually will be far less data than you would like to have available. Many kinds of rigorous analysis, such as intensive autecological studies and detailed PVAs, will not be possible in most cases. You will not know much about many of the target species and ecological processes of concern. Yet, plans must move forward regardless of these deficiencies. You and your fellow scientists will have to rely on professional judgment and general principles more than you might feel comfortable with initially.

2. Science should be the foundation for conservation planning, but don't expect planners to follow all of your suggestions. There are social, economic, and political factors that must be reconciled with biological concerns. On the other hand, remember it is your responsibility to assure that biology is recognized as the bottom line in planning decisions and that socioeconomic compromises do not threaten the scientific defensibility or biological viability of a plan.

3. Be as honest, objective, and unbiased as possible. You should not be ashamed to consider yourself a conservationist, as well as a scientist, but it is your primary responsibility to advocate a strict adherence to science throughout the planning process. The minute you step over the line from responsible advocacy of science and adherence to accepted conservation goals, to advocacy of particular interests and positions, you risk being marginalized and your participation in the process being permanently impaired.

4. Do not let your science be corrupted or used improperly by any parties in the negotiations. Expect that environmentalists and developers (or industry) will both try to use your name or words to support their interests, and they may grossly distort what you actually said or wrote. Expect also that some environmentalists or developers will attack you viciously and publicly if you do not say everything they want to hear. Try your best to not be used or abused in these ways, but be "thick-skinned" and able to ignore what you cannot change.

For Citizen Activists

1. Recognize that habitat-based conservation planning, whether you like all aspects of it or not, is here to stay. For the foreseeable future HCPs, NCCPs, and similar planning exercises will be on the rise in many regions of the United States. Trying to stop or undermine these plans is counterproductive. Instead, become an informed and responsible participant and try to make plans as biologically conservative as possible.

2. Your role in conservation planning is to advocate protection and restoration of the native biota and other environmental values. You can play this role while still being honest, courteous, and fair to other participants. You will not be taken seriously if you appear extremist, dishonest, rude, or "flaky." On the other hand, it is your responsibility to fight compromises that put native species and ecosystems at high risk. Even modest conservation proposals will be painted as radical and extremist by some prodevelopment interests. Therefore, don't sacrifice essential values in an effort to appear reasonable.

3. Try to appreciate the proper role of scientists in conservation planning. They should not be employed as hired guns and you will not like everything they have to say. Do not attack them when they do not follow your party line. You will be more respected and more influential in the planning process if you serve as honest watchdogs of science and sincere advocates for widely shared conservation values.

4. Understand that the use of habitat-based conservation planning to prevent listings of species under the ESA has benefits and

dangers both. Do not automatically oppose attempts to prevent listings. A proper purpose of conservation planning, thoroughly in line with the goal of the ESA to conserve ecosystems, is to maintain habitats in such a condition that species do not decline to the point where they require listing. A good conservation plan should be able to prevent listings. On the other hand, many species will require the individual attention afforded by the ESA until their habitats can be protected or restored to the point where their recovery is assured. Without the regulatory "teeth" of the ESA for at least some species, conservation plans are unlikely to progress. Make these distinctions carefully.

For Developers and Industry

1. Do not assume that conservationists are your enemy. Property values generally increase as more of a landscape is protected as natural area and as the quality of life for all residents improves. By being open-minded to conservation you gain positive public relations and, ultimately, more political influence.

2. You have a right to expect assurances that you will not have to come back continually to the planning table or foot the bill for changes in conservation plans. Conservation benefits the public, and the public should pay a fair share of the bill. On the other hand do not expect plans to be immutable. Adaptive management is an accepted paradigm for conservation planning and it requires that plans be revised on the basis of new information. So, although you should not be expected to pay directly for revisions needed because of factors outside of your control, there will be costs, and your development plans should be flexible enough to accommodate change.

3. Support a strong foundation in objective science for conservation plans. Do not try to erode the objectivity of scientists in any way. A conservation plan will be more defensible and your interests ultimately more secure if scientists proceed independently from influences from either side.

4. You are among many diverse interests involved in a conservation plan. Do not expect your wealth and political influence to carry the day in a process that is ultimately democratic. Closed-door

deals may have worked in the past, in some cases, but conservation planning is becoming more open to public scrutiny. With the
public assuming the risk of plan failures and the expense of adaptive management, they have even more of a right to be intimately
involved in the planning process.

For Agencies

1. You have the responsibility for administering conservation plans.
   As representatives of the public interest you must ensure wide
   public participation. Your commitment to the democracy of the
   process is essential. Nevertheless, the biggest mistake would be to
   consider each voice at the planning table as equal to all others.
   Some voices are simply more honest, more informed, and more
   altruistic than others. Strict adherence to biologically credible and
   widely (but not necessarily unanimously) agreed-on conservation goals is the best way to avoid subversion of the process by
   special interests on either side.

2. As agencies you are especially subject to political influence, which
   you must strive to avoid as much as possible. The best way to re-
   sist adverse political influence is, as stated above, to adhere to
   goals and to recognize science as the foundation of the process.
   Look to scientists and their professional societies for support, ideally before the political pressure becomes overbearing.

3. Bureaucratic red tape and delays in funding and implementing research and other phases of planning are among the major causes
   of failures of conservation plans. Although it will not be easy, find
   ways to streamline proposal solicitations and reviews, contracting, research permitting, and implementation of plan components. Concomitantly, find ways to reduce bureaucratic costs.
   Even with the best science available and abundant political good
   will, the conservation planning process will fail if bureaucracies
   lack the efficiency to see it through.

4. Be careful about how you "sell" conservation planning to the
   public. The argument for avoiding listings of species under the
   ESA carries many dangers. Discriminate carefully those species
   whose listing might be legitimately precluded by conservation
   planning from those that require listing in order to recover, and

whose listing will provide the needed incentive for private parties to participate in planning.

If the parties involved in conservation planning are willing to take these suggestions, or at least a good number of them, we are confident that many of the present conflicts over HCPs and other plans can be avoided in future plans. Most important, we believe that the species and ecosystems affected by plans will be better off. This, after all, is why we have an ESA.

As we conclude this book Congress is once again poised to consider competing bills to reauthorize the ESA. There will be much argument between industry and environmental interests, and various splits will occur within both camps. By the time this book is on the bestseller's list (just checking to see if you're paying attention, reader) a reauthorization bill may have passed, and there will be new incentives for implementing much of what we have recommended here, as well as new constraints on achieving any of our proposals. We feel that the basic principles and approach offered here will be valid regardless of changes in the law. We predict that habitat-based conservation planning, under the ESA and otherwise, will blossom in coming years and decades. It can either proceed the way it largely has to date—absent meaningful scientific input and adaptive management in many cases and without adequate mitigation for habitat loss—or it can improve and actually contribute to the recovery of listed species and the ecosystems on which they depend. It seems a simple choice.

LITERATURE CITED

Acevedo, M. F., D. L. Urban, and M. Ablan. 1996. Landscape scale forest dynamics: GIS, gap, and transition models. Pp. 181–185 in M. F. Goodchild, L. T. Steyaert, B. O. Parks, C. Johnston, D. Maidment, M. Crane, and S. Glendinning, eds., *GIS and Environmental Modeling: Progress and Research Issues.* GIS World Books, Ft. Collins, CO.

Adkison, G. P., and M. T. Jackson. 1996. Changes in ground-layer vegetation near trails in midwestern U.S. forests. *Natural Areas Journal* 16:14–23.

Akçakaya, H. R., and J. L. Atwood. 1997. A habitat-based metapopulation model of the California gnatcatcher. *Conservation Biology* 11:422–434.

Allan, J. D., and A. S. Flecker. 1993. Biodiversity conservation in running waters. *Bio-Science* 43:32–43.

Anderson, T. L., and J. J. Olsen. 1993. Positive incentives for saving endangered species. Pp. 109–114 in H. Fischer and W. E. Hudson, eds., *Building Economic Incentives into the Endangered Species Act.* Defenders of Wildlife, Washington, DC.

Andrewartha, H. G., and L. C. Birch. 1984. *The Ecological Web: More on the Distribution and Abundance of Animals.* University of Chicago Press, Chicago, IL.

Angermeier, P. L., and J. R. Karr. 1994. Biological integrity versus biological diversity as policy directives. *BioScience* 44:690–697.

Anstett, M.-C., M. Hossaert-McKey, and D. McKey. 1997. Modeling the persistence of small populations of strongly interdependent species: Figs and fig wasps. *Conservation Biology* 11: 204–213.

Atwood, J. L. 1993. California gnatcatchers and coastal sage scrub: The biological basis for endangered species listing. Pp. 149–169 in J. E. Keeley, ed., *Interface between ecology and land development in California.* Southern California Academy of Sciences, Los Angeles.

Atwood, J. L., and R. F. Noss. 1994. Gnatcatchers and development: A "train wreck" avoided? *Illahee* 10:123–130.

Austin, D. F. 1976. Florida scrub. *Florida Naturalist* 49:2–5.

Baden, J. A. and T. O'Brien. 1993. Toward a true ESA: An ecological stewardship act. Pp. 95–100 in H. Fischer and W. E. Hudson, eds., *Building economic incentives into the endangered species act.* Defenders of Wildlife, Washington, DC.

Bailey, J. A. 1982. Implications of "muddling through" for wildlife management. *Wildlife Society Bulletin* 10:363–369.

Barrows, C. W. 1996. An ecological model for the protection of a dune ecosystem. *Conservation Biology* 10:888–891.

Barry, D. J. 1991. The interagency consultation process under section 7 of the Endangered Species Act. Testimony before the Senate Environment and Public Works Committee. April 10. Washington, DC.

Barry, D., and M. Oelschlaeger. 1996. A science for survival: Values and conservation biology. *Conservation Biology* 10:905–911.

Bayne, E. M., and K. A. Hobson. 1997. Comparing the effects of landscape fragmentation by forestry and agriculture on predation of artificial nests. *Conservation Biology* 11 (in press).

Bean, M. J. 1983. *The evolution of national wildlife law.* Praeger Publishing, New York, NY.

Bean, M. J., S. G. Fitzgerald, and M. A. O'Connell. 1991. *Reconciling conflicts under the Endangered Species Act: the habitat conservation planning experience.* World Wildlife Fund, Washington, DC.

Beatley, T. 1994. *Habitat conservation planning: endangered species and urban growth.* University of Texas Press, Austin, TX.

Bedward, M., R. L. Pressey, and D. A. Keith. 1992. A new approach for selecting fully representative reserve networks: Addressing efficiency, reserve design, and land suitability with an iterative analysis. *Biological Conservation* 62:115–125.

Bennett, A. F. 1990. *Habitat Corridors: Their Role in Wildlife Management and Conservation.* Arthur Rylah Institute for Environmental Research, Department of Conservation and Environment, Victoria, Australia.

Bingham, B. B., and B. R. Noon. 1997. Mitigation of habitat "take": Application to habitat conservation planning. *Conservation Biology* 11:127–139.

Boyce, M. S. 1992. Population viability analysis. *Annual Review of Ecology and Systematics* 23:481–506.

Brewer, G. D., and T. W. Clark. 1994. A policy sciences perspective: Improving implementation. Pp. 391–413 in T. W. Clark, R. P. Reading, and A. L. Clarke, eds. *Endangered Species Recovery: finding the lessons, improving the process.* Island Press, Washington, DC.

Brown, J. H., and A. Kodric-Brown. 1977. Turnover rates in insular biogeography: Effect of immigration on extinction. *Ecology* 58:445–449.

Brunner, R. D., and T. W. Clark. 1997. A practice-based approach to ecosystem management. *Conservation Biology* 11:48–58.

Burgess R. L., and D. M. Sharpe, eds. 1981. *Forest island dynamics in man-dominated landscapes.* Springer-Verlag, New York, NY.

California Department of Fish and Game (CDFG). 1993. Natural community conservation planning: Conservation guidelines. Unpublished document. California Department of Fish and Game, Sacramento, CA.

Callicott, J. B., and K. Mumford. 1997. Ecological sustainability as a conservation concept. *Conservation Biology* 11:32–40.

Capra, F. 1975. *The Tao of Physics.* Shambhala, Boulder, CO.

Christensen, N. L., A. M. Bartuska, J. H. Brown, S. Carpenter, C. D'Antonio, R. Francis, J. F. Franklin, J. A. MacMahon, R. F. Noss, D. J. Parsons, C. H. Peterson, M. G. Turner, and R. G. Woodmansee. 1996. The report of the Ecological Society of America Committee on the Scientific Basis for Ecosystem Management. *Ecological Applications* 6:665–691.

Christman, S. P., and W. S. Judd. 1990. Notes on plants endemic to Florida scrub. *Florida Scientist:* 53 (1): 52–73.

Clark, T. W., G. N. Backhouse, and R. P. Reading. 1995. Prototyping in endangered species recovery programmes: The eastern barred bandicoot experience. Pp. 50–62 in A. Bennett, G. N. Backhouse, and T. W. Clark, eds., *People and Nature conservation: perspectives on private land use and endangered species recovery.* Transactions of the Royal Society of New South Wales, Chipping Norton, Australia.

Clark, T. W., P. C. Paquet, and A. P. Curlee. 1996. General lessons and positive trends in large carnivore conservation. *Conservation Biology* 10:1055–1058.

Costanza, R. 1993. Developing ecological research that is relevant for achieving sustainability. *Ecological Applications* 3:579–581.

County of Orange Environmental Management Agency. 1996. *Natural community conservation plan and habitat conservation plan: environmental documentation.* 4 vols. Board of Supervisors of the County of Orange, CA.

Cox, J. A. 1987. *Status and Distribution of the Florida Scrub Jay.* Florida Ornithological Society Special Publication No. 3, 110 pp.

Cox, J., D. Inkley, and R. Kautz. 1987. Ecology and habitat protection needs of gopher tortoise *(Gopherus polyphemus)* populations found on lands slated for large-scale development in Florida. Florida Game and Fresh Water Fish Commission Non-game Wildlife Program Technical Report No. 4.

Cox, J., R. Kautz, M. MacLauglin, and T. Gilbert. 1994. Closing the gaps in Florida's wildlife habitat conservation system. Florida Game and Fresh Water Fish Commission, Tallahassee, FL.

Darwin, C. R. 1859. *On the Origin of Species by Means of Natural Selection.* John Murray, London.

DellaSala, D., J. R. Strittholt, R. F. Noss, and D. M. Olson. 1996. A critical role for core reserves in managing Inland Northwest landscapes for natural resources and biodiversity. *Wildlife Society Bulletin* 24:209–221.

Diamond, J. M. 1975. The island dilemma: Lessons of modern biogeographic studies for the design of natural preserves. *Biological Conservation* 7: 129–146.

———. 1976. Island biogeography and conservation: Strategy and limitations. *Science* 193:1027–1029.

———. 1984. Historic extinctions: A Rosetta stone for understanding prehistoric extinctions. Pp. 824–862 in P. S. Martin and R. G. Klein, eds., *Quaternary Extinctions: A Prehistoric Revolution.* University of Arizona Press, Tucson, AZ.

———. 1986. Overview: Laboratory experiments, field experiments, and natural experiments. Pp. 3–22 in J. Diamond and T. J. Case, eds., *Community Ecology.* Harper and Row, New York.

Dinerstein, E., D. M. Olson, D. J. Graham, A. L. Webster, S. A. Primm, M. P. Bookbinder, and G. Ledec. 1995. *A conservation assessment of the terrestrial ecoregions of Latin America and the Caribbean.* World Wildlife Fund, Washington, DC.

Doak, D. 1989. Spotted owls and old growth logging in the Pacific Northwest. *Conservation Biology* 3:389–396.

———. 1995. Source-sink models and the problem of habitat degradation: General models and applications to the Yellowstone grizzly. *Conservation Biology* 9:1370–1379.

Dobson, A. P., J. P. Rodriguez, W. M. Roberts, and D. S. Wilcove. 1997. Geographic distribution of endangered species in the United States. *Science* 275:550–553.

Drayton, B., and R. B. Primack. 1996. Plant species lost in an isolated conservation area in Metropolitan Boston from 1894 to 1993. *Conservation Biology* 10:30–39.

Duever, L. C., R. W. Simons, R. F. Noss, and J. R. Newman. 1987. *Comprehensive inventory of natural ecological communities in Alachua County.* KBN Engineering and Applied Sciences, Gainesville, FL.

Dunning, J. B., B. J. Danielson, and H. R. Pulliam. 1992. Ecological processes that affect populations in complex landscapes. *Oikos* 65:169–175.

Dunning, J. B., D. J. Stewart, B. J. Danielson, B. R. Noon, T. L. Root, R. H. Lamberson, and E. E. Stevens. 1995. Spatially explicit population models: Current forms and future uses. *Ecological Applications* 5:3–11.

Dwyer, L. E., D. D. Murphy, and P. R. Ehrlich. 1995a. Property rights case law and the challenge to the Endangered Species Act. *Conservation Biology* 9:725–741.

Dwyer, L. E., D. D. Murphy, S. P. Johnson, and M. A. O'Connell. 1995b. Avoiding the trainwreck: Observations from the frontlines of natural community conservation planning in southern California. *Endangered Species Update* 12 (12): 5–7.

Ehrenfeld, D. 1978. *The Arrogance of Humanism.* Oxford University Press, New York.

Ehrlich, P. R., and A. H. Ehrlich. 1981. *Extinction: The Causes and Consequences of the Disappearance of Species.* Random House, New York.

———. 1996. *Betrayal of Science and Reason: How Anti-Environmental Rhetoric Threatens Our Future.* Island Press, Washington, DC.

Ehrlich, P. R., and E. O. Wilson. 1991. Biodiversity studies: Science and policy. *Science* 253:758–762.

Fahrig, L., and G. Merriam. 1994. Conservation of fragmented populations. *Conservation Biology* 8:50–59.

Fischer, H., and W. E. Hudson, eds. 1993. *Building Economic Incentives into the Endangered Species Act.* Defenders of Wildlife, Washington, DC.

Fitzpatrick, J. W., G. E. Woolfenden, and M. T. Kopeny. 1991. Ecology and development related habitat requirements of the Florida scrub jay *(Aphelocoma coerulescens coerulescens).* Florida Game and Fresh Water Fish Commission Non-game Wildlife Program Technical Report No. 8.

Fitzpatrick, J. W., R. Bowman, D. R. Breininger, M. A. O'Connell, B. Stith, J. Thaxton, B. Toland, and G. W. Woolfenden. 1997. *Habitat conservation plans for the Florida scrub jay (Aphelocoma c. coerulescens): A Biological Framework.* American Ornithologists' Union, Washington, DC.

Flather, C. H., L. A. Joyce, and C. A. Bloomgarden. 1994. Species endangerment patterns in the United States. USDA Forest Service Technical Report RM-241. Fort Collins, CO.

Frankel, O. H., and M. E. Soulé. 1981. *Conservation and Evolution.* Cambridge University Press. Cambridge, UK.

Franklin, I. R. 1980. Evolutionary changes in small populations. Pp. 135–149 in M. E. Soulé and B. A. Wilcox, eds., *Conservation Biology: An Evolutionary-Ecological Perspective.* Sinauer Associates, Sunderland, MA.

Fredrick, P. C., K. L. Bildstein, B. Fleury, and J. Ogden. 1996. Conservation of large, nomadic populations of white ibises *(Eudocimus albus)* in the United States. *Conservation Biology* 10:203–216.

Friesen, L. E., P. F. J. Eagles, and R. J. MacKay. 1995. Effects of residential development on forest-dwelling neotropical migrant songbirds. *Conservation Biology* 9:1408–1414.

Gentry, A. W. 1986. Endemism in tropical versus temperate plant communities. Pp. 153–181 in M. E. Soulé, ed., *Conservation Biology: The Science of Scarcity and Diversity.* Sinauer Associates, Sunderland, MA.

Gilligan, D. M., L. M. Woodworth, M. E. Montgomery, David A. Briscoe, and

R. Frankham. 1997. Is mutation accumulation a threat to the survival of endangered populations? *Conservation Biology* 11 (in press).

Gilpin, M. E., and I. Hanski, eds. 1991. *Metapopulaton Dynamics: Empirical and Theoretical Investigations.* Linnaean Society of London and Academic Press, London.

Gilpin, M. E., and M. E. Soulé. 1986. Minimum viable populations: Processes of species extinction. Pp. 19–34 in M. E. Soulé, ed., *Conservation Biology: The Science of Scarcity and Diversity.* Sinauer Associates, Sunderland, MA.

Givens, K. T., J. N. Layne, W. G. Abrahamson, and S. C. White-Schuler. 1984. Structural changes and successional relationships of five Florida Lake Wales Ridge plant communities. *Bulletin Torrey Botanical Club* 111:8–18.

Goodman, D. 1987. The demography of chance extinction. Pp. 11–34 in M. E. Soulé, ed., *Viable Populations for Conservation.* Cambridge University Press, Cambridge, UK.

Grove, R. H. 1992. Origins of western environmentalism. *Scientific American,* July, 42–47.

Hanski, I., A. Moilanen, T. Pakkala, and M. Kuussaari. 1996. The quantitative incidence function model and persistence of an endangered butterfly metapopulation. *Conservation Biology* 10:578–590.

Harris, L. D. 1984. *The Fragmented Forest: Island Biogeography Theory and the Preservation of Biotic Diversity.* University of Chicago Press, Chicago, IL.

Harrison, S. 1994. Metapopulations and conservation. Pp. 111–128 in P. J. Edwards, R. M. May, and N. R. Webb, eds., *Large-Scale Ecology and Conservation Biology.* Blackwell Science, Oxford, UK.

Herkert, J. R. 1994. The effects of habitat fragmentation on midwestern grassland bird communities. *Ecological Applications* 4:461–471.

Hobbs, R. J. 1992. The role of corridors in conservation: Solution or bandwagon? *Trends in Ecology and Evolution* 7:389–392.

Holling, C. S., ed. 1978. *Adaptive Environmental Assessment and Management.* John Wiley and Sons, New York.

Holling, C. S., and G. K. Meffe. 1996. Command and control and the pathology of natural resource management. *Conservation Biology* 10:328–337.

Host, G. E., P. Polzer, D. J. Mladenoff, M. A. White, and T. R. Crow. 1996. A quantitative approach to developing regional ecosystem classifications. *Ecological Applications* 6:608–618.

Houck, O. A. 1993. The Endangered Species Act and its implementation by the U. S. Departments of Interior and Commerce. *University of Colorado Law Review* 64 (2).

———. 1995. Why do we protect endangered species, and what does that say about whether restrictions on private property to protect them constitute "takings"? *Iowa Law Review* 80:297–332.

Hough, J. 1988. Biosphere reserves: Myth and reality. *Endangered Species Update* 6 (1&2): 1–4.

Hurlbert, S. H. 1984. Pseudoreplication and the design of ecological field experiments. *Ecological Monographs* 54:187–211.

Hutto, R. L., S. Reel, and P. B. Landres. 1987. A critical evaluation of the species approach to biological conservation. *Endangered Species Update* 4 (12): 1–4.

Irvine, S. 1994. The cornucopia scam: Contradictions of sustainable development. Part 1: Ignoring the limits to growth. *Wild Earth* 4 (3): 73–81.

IUCN Species Survival Commission. 1994. *IUCN Red List Categories.* IUCN, Gland, Switzerland.

Jackson, D. R., and E. G. Milstrey. 1989. The fauna of gopher tortoise burrows. Pp. 86–98 in J. E. Diemer, D. R. Jackson, J. L. Landers, J. N. Layne, and D. A. Wood, eds., *Gopher Tortoise Relocation Symposium Proceedings.* Nongame Wildlife Program Technical Report No. 5. Florida Game and Fresh Water Fish Commission, Tallahassee, FL.

James, F. C., C. A. Hess, and D. Kufrin. 1997. Species-centered environmental analysis: Indirect effects of fire history on red-cockaded woodpeckers. *Ecological Applications* 7:118–129.

Jost, K. 1996. Protecting endangered species. *The CQ Researcher* 6 (15): 337–360. Congressional Quarterly, Washington, DC.

Karr, J. R. 1981. Assessment of biotic integrity using fish communities. *Fisheries* 6:21–27.

———. 1982. Population variability and extinction in the avifauna of a tropical land bridge island. *Ecology* 63:1975–1978.

———. 1991. Biological integrity: A long-neglected aspect of water resource management. *Ecological Applications* 1:66–84.

Keddy, P. A., H. T. Lee, and I. C. Wisheu. 1993. Choosing indicators of ecosystem integrity: Wetlands as a model system. Pp. 61–79 in S. Woodley, J. Kay, and G. Francis, eds., *Ecological Integrity and the Management of Ecosystems.* St. Lucie Press, Ottawa, Canada.

Keiter, R. B. 1994. Conservation biology and the law: Assessing the challenges ahead. *Chicago-Kent Law Review* 69:911–933.

Keystone Policy Center. 1995. *Keystone Dialogue on Incentives to Protect Endangered Species on Private Lands: Final Report.* Keystone Center, Keystone, CO.

Kiester, A. R., J. M. Scott, B. Csuti, R. F. Noss, B. Butterfield, K. Sahr, and D. White. 1996. Conservation prioritization using GAP data. *Conservation Biology* 10:1332–1342.

Kuhn, T. S. 1970. *The Structure of Scientific Revolutions.* 2d ed. University of Chicago Press, Chicago.

Laessle, A. M. 1958. The origin and successional relationships of sandhill vegetation and sand pine scrub. *Ecological Monographs* 28:361–387.

———. 1968. Relationships of sand pine scrub to former shore lines. *Quarterly Journal of the Florida Academy of Sciences* 30:269–286.

Lambeck, R. J. 1997. Focal species define landscape requirements for nature conservation. *Conservation Biology* 11:849–856.

Lamberson, R. H., R. McKelvey, B. R. Noon, and C. Voss. 1992. A dynamic analysis of northern spotted owl viability in a fragmented forest landscape. *Conservation Biology* 6:505–512.

Lande, R. 1988. Demographic models of the northern spotted owl *(Strix occidentalis caurina). Oecologia* 75:601–607.

———. 1995. Mutation and conservation. *Conservation Biology* 9:782–791.

Lande, R., and G. F. Barrowclough. 1987. Effective population size, genetic variation, and their use in population management. Pp. 87–123 in M. E. Soulé, ed., *Viable Populations for Conservation.* Cambridge University Press, Cambridge, UK.

Leopold, A. 1933. *Game management.* Charles Scribner's Sons, New York.

———. 1941. Wilderness as a land laboratory. *Living Wilderness* 6 (July): 3.

———. 1949. A *Sand County Almanac.* Oxford University Press, New York.

———. 1953. *Round River.* Oxford University Press, New York.

Lesica, P., and F. W. Allendorf. 1995. When are peripheral populations valuable for conservation? *Conservation Biology* 9:753–760.

Liebig, J. 1840. *Chemistry and Its Application to Agriculture and Physiology.* Taylor and Walton, London.

Lindenmayer, D. B., and H. P. Possingham. 1996. Ranking conservation and timber management options for Leadbeater's possum in southeastern Australia using population viability analysis. *Conservation Biology* 10:235–251.

Lomolino, M. V., and R. Channell. 1995. Splendid isolation: Patterns of geographic range collapse in endangered mammals. *Journal of Mammalogy* 76:335–347.

Ludwig, D., R. Hilborn, and C. Walters. 1993. Uncertainty, resource exploitation, and conservation: Lessons from history. *Science* 260:17, 36.

Lyster, S. 1985. *International Wildlife Law.* Grotius Publications, Llandysul, Dyfed, UK.

MacArthur, R. H. 1972. *Geographical Ecology: Patterns in the Distribution of Species.* Princeton University Press, Princeton, NJ.

MacArthur, R. H., and E. O. Wilson. 1967. *The Theory of Island Biogeography.* Princeton University Press, Princeton, NJ.

McCarthy, M. A., M. A. Burgman, and S. Ferson. 1995. Sensitivity analysis for models of population viability. *Biological Conservation* 73:93–100.

McCaull, J. 1994. The natural community conservation planning program and the coastal sage scrub ecosystem of southern California. Pp. 281–292 in R. E. Grumbine, ed., *Environmental Policy and Biodiversity.* Island Press, Washington, DC.

McGarigal, K., and W. C. McComb. 1995. Relationships between landscape structure and breeding birds in the Oregon Coast Range. *Ecological Monographs* 65:235–260.

McKelvey, K., B. R. Noon, and R. H. Lamberson. 1993. Conservation planning for species occupying fragmented landscapes: The case of the northern spotted owl. Pp. 424–450 in P. M. Kareiva, J. G. Kingsolver, and R. B. Huey, eds., *Biotic Interactions and Global Change.* Sinauer Associates, Sunderland, MA.

McMillan, M. 1996. Effects of the moratorium on listings under the Endangered Species Act. *Endangered Species Update* 13 (3): 5–6.

Mace, G. M., and R. Lande. 1991. Assessing extinction threats: Toward a reevaluation of IUCN threatened species categories. *Conservation Biology* 5:148–157.

Maguire, L. A. 1994. Science, values, and uncertainty: A critique of The Wildlands Project. Pp. 267–279 in R. E. Grumbine, ed., *Environmental policy and biodiversity.* Island Press, Washington, DC.

———. 1996. Making the role of values in conservation explicit. *Conservation Biology* 10:914–916

Maidment, D. 1996. Environmental modeling within GIS. Pp. 315–323 in M. F. Goodchild, L. T. Steyaert, B. O. Parks, C. Johnston, D. Maidment, M. Crane, and S. Glendinning, eds., *GIS and Environmental Modeling: Progress and Research Issues.* GIS World Books, Ft. Collins, CO.

Mangel, M., and C. Tier. 1994. Four facts every conservation biologist should know about extinction. *Ecology* 75:607–614.

Martin, P. S., and R. G. Klein, eds. 1984. *Quaternary Extinctions: A Prehistoric Revolution.* University of Arizona Press, Tucson, AZ.

Mattson, D. J. 1996. Ethics and science in natural resource agencies. *BioScience* 46:767–771.

Mattson, D. J., S. Herrero, R. G. Wright, and C. M. Pease. 1996. Science and management of Rocky Mountain grizzly bears. *Conservation Biology* 10:1013–1025.

Mayr, E. 1963. *Animal Species and Evolution.* Belknap Press of Harvard University Press, Cambridge, MA.

Meffe, G. K., and C. R. Carroll, eds. 1994. *Principles of Conservation Biology.* Sinauer Associates, Sunderland, MA.

Merriam, G. 1991. Corridors and connectivity: Animal populations in heterogeneous environments. Pp. 133–142 in D. A. Saunders and R. J. Hobbs, eds., *Nature conservation 2: The role of corridors.* Surrey Beatty and Sons, Chipping Norton, NSW, Australia.

Miller, B., G. Ceballos, and R. Reading. 1994. The prairie dog and biotic diversity. *Conservation Biology* 8:677–681.

Miller, R. R., J. D. Williams, and J. E. Williams. 1989. Extinctions of North American fishes during the past century. *Fisheries* 14: 22–38.

Mills, L. S., S. G. Hayes, C. Baldwin, M. J. Wisdom, J. Citta, D. J. Mattson, and K. Murphy. 1996. Factors leading to different viability predictions for a grizzly bear data set. *Conservation Biology* 10:863–873.

Mladenoff, D. J., T. A. Sickley, R. G. Haight, and A. P. Wydeven. 1995. A regional landscape analysis and prediction of favorable gray wolf habitat in the northern Great Lakes region. *Conservation Biology* 9:279–294.

Mooney, H. A. 1988. Lessons from Mediterranean-climate regions. Pp. 157–165 in E. O. Wilson, ed., *Biodiversity.* National Academy Press, Washington, DC.

Murphy, D. D., and B. R. Noon. 1992. Integrating scientific methods with habitat conservation planning: Reserve design for northern spotted owls. *Ecological Applications* 2:3–17.

Murphy, D., D. Wilcove, R. Noss, J. Harte, C. Safina, J. Lubchenco, T. Root, V. Sher, L. Kaufman, M. Bean, and S. Pimm. 1994. On reauthorization of the Endangered Species Act. *Conservation Biology* 8:1–3.

Myers, R. L. 1985. Fire and the dynamic relationship between Florida sandhill and sand pine scrub vegetation. *Bulletin Torrey Botanical Club* 112:241–252.

Myers, R. L., and D. L. White. 1987. Landscape history and changes in sandhill vegetation in north-central and south-central Florida. *Bulletin Torrey Botanical Club* 114:21–32.

Nature Conservancy, The. 1992. Extinct vertebrate species in North America. Unpublished draft list, March 4, 1992. The Nature Conservancy, Arlington, VA.

Norton, T. W. 1995. Introduction (to special issue on population viability analysis). *Biological Conservation* 73:91.

Noss, R. F. 1983. A regional landscape approach to maintain diversity. *BioScience* 33: 700–706.

———. 1986. Dangerous simplifications in conservation biology. *Bulletin of the Ecological Society of America* 67: 278–279.

———. 1987a. From plant communities to landscapes in conservation inventories: A look at The Nature Conservancy (USA). *Biological Conservation* 41:11–37.

———. 1987b. Corridors in real landscapes: A reply to Simberloff and Cox. *Conservation Biology* 1:159–164.

———. 1987c. Protecting natural areas in fragmented landscapes. *Natural Areas Journal* 7:2–13.

———. 1990. Indicators for monitoring biodiversity: A hierarchical approach. *Conservation Biology* 4:355–364.

———. 1991. From endangered species to biodiversity. Pp. 227–246 in K. A. Kohm, ed., *Balancing on the Brink of Extinction: The Endangered Species Act and Lessons for the Future.* Island Press, Washington, DC.

———. 1992. The Wildlands Project: Land conservation strategy. *Wild Earth* (Special Issue): 10–25.

———. 1993. Wildlife corridors. Pp. 43–68 in D. S. Smith and P. C. Hellmund, eds., *Ecology of Greenways.* University of Minnesota Press, Minneapolis, MN.

———. 1994. Some principles of conservation biology, as they apply to environmental law. *Chicago-Kent Law Review* 69:893–909.

———. 1995. *Maintaining Ecological Integrity in Representative Reserve Networks.* World Wildlife Fund Canada, Toronto, Ontario.

———. 1996a. Ecosystems as conservation targets. *Trends in Ecology and Evolution* 11:351.

———. 1996b. Protected areas: How much is enough? Pp. 91–120 in R. G. Wright, ed., *National Parks and Protected Areas.* Blackwell, Cambridge, MA.

———. 1996c. The naturalists are dying off. *Conservation Biology* 10:1–3.

Noss, R. F., and A. Cooperrider. 1994. *Saving Nature's Legacy: Protecting and Restoring Biodiversity.* Defenders of Wildlife and Island Press, Washington, DC.

Noss, R. F., and B. Csuti. 1994. Habitat fragmentation. Pp. 237–264 in G. K. Meffe and R. C. Carroll, eds., *Principles of Conservation Biology.* Sinauer Associates, Sunderland, MA.

Noss, R. F., and L. D. Harris. 1986. Nodes, networks, and MUM's: Preserving diversity at all scales. *Environmental Management* 10:299–309.

Noss, R. F., and D. D. Murphy. 1995. Endangered species left homeless in Sweet Home. *Conservation Biology* 9:229–231.

Noss, R. F., and R. L. Peters. 1995. *Endangered Ecosystems of the United States: A Status Report and Plan for Action.* Defenders of Wildlife, Washington, DC.

Noss, R. F., E. T. LaRoe, and J. M. Scott. 1995. Endangered ecosystems of the United States: A preliminary assessment of loss and degradation. *Biological Report 28.* USDI National Biological Service, Washington, DC.

Noss, R. F., H. B. Quigley, M. G. Hornocker, T. Merrill, and P. C. Paquet. 1996. Conservation biology and carnivore conservation in the Rocky Mountains. *Conservation Biology* 10:949–963.

O'Connell, M. A., and S. P. Johnson. 1997. Improving conservation planning under the Endangered Species Act. *Endangered Species Update* 14 (1): 1–4.

O'Toole, R. 1990. Pay land managers to protect species. *The Environmental Forum: The Policy Journal of the Environmental Law Institute* 33:33–34.

Peterman, R. M. 1990a. Statistical power analysis can improve fisheries research and management. *Canadian Journal of Fisheries and Aquatic Science* 47:2–15.

———— 1990b. The importance of reporting statistical power: The forest decline and acid deposition example. *Ecology* 71:2024–2027.

Peters, R. L. 1996. Hope for the red-cockaded? *Defenders* 71 (4): 27–32.

Pickett, S.T.A., and J. N. Thompson. 1978. Patch dynamics and the design of nature reserves. *Biological Conservation* 13:27–37.

Pickett, S.T.A., and P. S. White. 1985. *The Ecology of Natural Disturbance and Patch Dynamics*. Academic Press, Orlando, FL.

Pickett, S. T. A., V. T. Parker, and P. L. Fiedler. 1992. The new paradigm in ecology: Implications for conservation biology above the species level. Pp. 65–88 in P. L. Fiedler and S. K. Jain, eds., *Conservation Biology: The Theory and Practice of Nature Conservation, Preservation, and Management*. Chapman and Hall, New York.

Pimm, S. L. 1993. Life on an intermittent edge. *Trends in Ecology and Evolution* 8:45–46.

Pimm, S. L., H. L. Jones, and J. Diamond. 1988. On the risk of extinction. *American Naturalist* 132:757–785.

Platt, W. J., G. W. Evans, and S. L. Rathbun. 1988. The population dynamics of a long-lived conifer *(Pinus palustris)*. *American Naturalist* 131:491–525.

Possingham, H. P., D. B. Lindenmayer, and T. W. Norton. 1993. A framework for improved management of threatened species based on population viability analysis (PVA). *Pacific Conservation Biology* 1:39–45.

Power, M. E., D. Tilman, J. A. Estes, B. A. Menge, W. J. Bond, L. S. Mills, G. Daily, J. C. Castilla, J. Lubchenco, and R. T. Paine. 1996. Challenges in the quest for keystones. *Conservation Biology* 46:609–620.

Prendergast, J. R., R. M. Quinn, J. H. Lawton, B. C. Eversham, and D. W. Gibbons. 1993. Rare species, the coincidence of diversity hotspots and conservation strategies. *Nature* 365:335–337.

Press, D., D. F. Doak, and P. Steinberg. 1996. The role of local government in the conservation of rare species. *Conservation Biology* 10:1538–1548.

Pressey, R. L., C. J. Humphries, C. R. Margules, R. I. Vane-Wright, and P. H. Williams. 1993. Beyond opportunism: Key principles for systematic reserve selection. *Trends in Ecology and Evolution* 8:124–128.

Primack, R. B. 1993. *Introduction to Conservation Biology*. Sinauer Associates, Sunderland, MA.

Quintana-Ascencio, P. F., and E. S. Menges. 1996. Inferring metapopulation dynamics from patch-level incidence of Florida scrub plants. *Conservation Biology* 10:1210–1219.

Redford, K. H., and S. E. Sanderson. 1992. The brief, barren marriage of biodiversity and sustainability? *Bulletin of the Ecological Society of America* 73:36–39.

Ricketts, T., E. Dinerstein, D. M. Olson, C. Loucks, P. Hedao, K. Carney, S. Walters, and P. Hurley. 1997. *A Conservation Assessment of Terrestrial Ecoregions of North America*. World Wildlife Fund, Conservation Science Program, Washington, DC.

Robbins, C. S., D. K. Dawson, and B. A. Dowell. 1989. Habitat area requirements of breeding forest birds of the Middle Atlantic states. *Wildlife Monographs* 103:1–34.

Robinson, G. R., M. E. Yurlina, and S. N. Handel. 1994. A century of change in the Staten Island flora: Ecological correlates of species losses and invasions. *Bulletin of the Torrey Botanical Club* 121:119–129.

Robinson, J. G. 1993. The limits to caring: Sustainable living and the loss of biodiversity. *Conservation Biology* 7:20–28.

Ruckelshaus, M., C. Hartway, and P. Kareiva. 1997. Assessing the data requirements of spatially explicit dispersal models. *Conservation Biology* 11 (in press).

Runte, A. 1987. *National parks: The American experience.* 2d ed. University of Nebraska Press, Lincoln, NB.

Russell, C., and L. Morse. 1992. Extinct and possibly extinct plant species of the United States and Canada. Unpublished report. Review draft, 13 March 1992. The Nature Conservancy, Arlington, VA.

Ryti, R. T. 1992. Effect of the focal taxon on the selection of nature reserves. *Ecological Applications* 2:404–410.

Saetersdal, M., J. M. Line, and H.J.B. Birks. 1993. How to maximize biological diversity in nature reserve selection: Vascular plants and breeding birds in deciduous woodlands, western Norway. *Biological Conservation* 66:131–138.

Saunders, D. A., R. J. Hobbs, and C. R. Margules. 1991. Biological consequences of ecosystem fragmentation: A review. *Conservation Biology* 5:18–32.

Schemske, D. W., B. C. Husband, M. H. Ruckelhaus, C. Goodwillie, I. M. Parker, and J. G. Bishop. 1994. Evaluating approaches to the conservation of rare and endangered plants. *Ecology* 75:584–606.

Schiffman, P. M. 1994. Promotion of exotic weed establishment by endangered giant kangaroo rats in a California grassland. *Biodiversity and Conservation* 3:524–537.

Schumaker, N. 1996. Using landscape indices to predict habitat connectivity. *Ecology* 77:1210–1225.

Scientific Review Panel, Southern California Coastal Sage Scrub. 1992. Coastal sage scrub survey guidelines. California Department of Fish and Game, Sacramento, CA.

Scott, J. M., B. Csuti, J. D. Jacobi, and J. E. Estes. 1987. Species richness: A geographic approach to protecting future biological diversity. *BioScience* 37:782–788.

Scott, J. M., B. Csuti, K. Smith, J. E. Estes, and S. Caicco. 1991a. Gap analysis of species richness and vegetation cover: An integrated biodiversity conservation strategy. Pp. 282–297 in K. A. Kohm, ed., *Balancing on the Brink of Extinction: The Endangered Species Act and Lessons for the Future.* Island Press, Washington, DC.

Scott, J. M., B. Csuti, and S. Caicco. 1991b. Gap analysis: Assessing protection needs. Pp. 15–26 in W. E. Hudson, ed., *Landscape Linkages and Biodiversity.* Defenders of Wildlife and Island Press, Washington, DC.

Scott, J. M., F. Davis, B. Csuti, R. Noss, B. Butterfield, C. Groves, J. Anderson, S. Caicco, F. D'Erchia, T. C. Edwards, J. Ulliman, and R. G. Wright. 1993. Gap analysis: A geographical approach to protection of biological diversity. *Wildlife Monographs* 123:1–41.

Shaffer, M. L. 1981. Minimum population sizes for species conservation. *BioScience* 31:131–134.

———. 1992. *Keeping the Grizzly Bear in the American West: A Strategy for Real Recovery.* The Wilderness Society, Washington, DC.

Shelford, V. E. 1913. *Animal Communities in Temperate America.* University of Chicago Press, Chicago.

Shrader-Frechette, K. 1996. Throwing out the bathwater of positivism, keeping the baby of objectivity: Relativism and advocacy in conservation biology. *Conservation Biology* 10:912–914.

Shrader-Frechette, K. S., and E. D. McCoy. 1993. *Method in ecology: Strategies for conservation.* Cambridge University Press, Cambridge, UK.

Simberloff, D., and L. G. Abele. 1976. Island biogeography theory and conservation practice. *Science* 191:285–286.

Simberloff, D., and J. Cox. 1987. Consequences and costs of conservation corridors. *Conservation Biology* 1:63–71.

Simberloff, D., J. A. Farr, J. Cox, and D. W. Mehlman. 1992. Movement corridors: Conservation bargains or poor investments? *Conservation Biology* 6:493–504.

Smith, R. E. 1974. *Ecology and Field Biology.* 2d ed. Harper and Row, New York.

Snodgrass, J. W., T. Townsend, P. Brabitz, M. Chitwood, and D. Jordan. 1991. An inventory of scrub habitat and Florida scrub jays in Brevard County, Florida. Brevard County Office of Natural Resources Management, Melbourne, FL.

Soulé, M. E. 1980. Thresholds for survival: Maintaining fitness and evolutionary potential. Pp. 151–170 in M. E. Soulé and B. A. Wilcox, eds. *Conservation Biology: An Evolutionary-Ecological Perspective.* Sinauer Associates, Sunderland, MA.

———. 1985. What is conservation biology? *Bioscience* 35:727–734.

———. 1987. Where do we go from here? Pp. 175–183 in M. E. Soulé, ed., *Viable Populations for Conservation.* Cambridge University Press, Cambridge, UK.

———. 1991. Land use planning and wildlife maintenance: Guidelines for conserving wildlife in an urban landscape. *Journal of the American Planning Association* 57:313–323.

———. 1996. Are ecosystem processes enough? *Wild Earth* 6 (1): 59–60.

Soulé, M. E., and D. Simberloff. 1986. What do genetics and ecology tell us about the design of nature reserves? *Biological Conservation* 35:19–40.

Soulé, M. E., and B. A. Wilcox. 1980. Conservation biology: Its scope and challenge. Pp. 1–8 in M. E. Soulé and B. A. Wilcox, eds., *Conservation Biology: An Evolutionary-Ecological Perspective.* Sinauer Associates, Sunderland, MA.

Stanley, T. R. 1995. Ecosystem management and the arrogance of humanism. *Conservation Biology* 9:255–262.

Steadman, D. W. 1995. Prehistoric extinctions of Pacific island birds: Biodiversity meets zooarchaeology. *Science* 267:1123–1131.

Strittholt, J. R., and R.E.J. Boerner. 1995. Applying biodiversity gap analysis in a regional nature reserve design for the Edge of Appalachia, Ohio (USA). *Conservation Biology* 9:1492–1505.

Stroup, R. 1992. Takings and environmental habitat. In *Drawing the Line: Property Rights and Environmental Protection.* Washington Research Council/Center for Competitive Strategies, Olympia, WA.

Taylor, B. L., and T. Gerrodette. 1993. The uses of statistical power in conservation biology: The vaquita and northern spotted owl. *Conservation Biology* 7:489–500.

Tear, T. H., J. M. Scott, P. H. Hayward, and B. Griffith. 1993. Status and prospects for success of the Endangered Species Act: A look at recovery plans. *Science* 262:976–977.

———. 1995. Recovery plans and the Endangered Species Act: Are criticisms supported by data? *Conservation Biology* 9:182–195.

Terborgh, J., and B. Winter. 1983. A method for siting parks and reserves with special reference to Columbia and Ecuador. *Biological Conservation* 27:45–58.

Thomas, C. D. 1990. What do real population dynamics tell us about minimum viable population sizes? *Conservation Biology* 4:324–327.

Thomas, J. W., E. D. Forsman, J. B. Lint, E. C. Meslow, B. R. Noon, and J. Verner. 1990.

A conservation strategy for the northern spotted owl. USDA Forest Service, USDI Bureau of Land Management, USDI Fish and Wildlife Service, and USDI National Park Service, Portland, OR.

Thornton, R. D. 1991. Searching for consensus and predictability: Habitat conservation planning under the Endangered Species Act of 1973. *Environmental Law Review*. Northwestern School of Law of Lewis and Clark College, Portland, OR.

Tomlin, C. D., and K. M. Johnston. 1990. An experiment in land-use allocation with a geographic information system. Pp. 159–169 in D. J. Peuquet and D. F. Marble, eds., *Introductory Readings in Geographic Information Systems*. Taylor and Francis, New York.

Tscharntke, T. 1992. Fragmentation of *Phragmites* habitats, minimum viable population size, habitat suitability, and local extinction of moths, midges, flies, aphids, and birds. *Conservation Biology* 6:530–536.

Turner, M. G., and R. H. Gardner. 1991. *Quantitative Methods in Landscape Ecology: The Analysis and Interpretation of Landscape Heterogeneity*. Springer-Verlag, New York.

UNESCO. 1974. Task force on criteria and guidelines for the choice and establishment of biosphere reserves. *Man and the Biosphere Report*, No. 22. Paris, France.

U. S. Fish and Wildlife Service. 1993. Endangered and threatened wildlife and plants: Threatened coastal California gnatcatcher. Final rule and proposed special rule. *Federal Register* 58:16742–16758.

———. 1994. Report to Congress. Recovery program, endangered and threatened species. U. S. Fish and Wildlife Service, Washington, DC.

———. 1995. HCPs in development per region by size category as of September 1, 1995. Division of Endangered Species. Washington, DC.

———. 1996. Endangered and threatened species list. Division of Endangered Species. February 19. Washington, DC.

U. S. General Accounting Office. 1994. Endangered species act: Information on species protection on nonfederal lands. GAO/RCED-95-16. Washington, DC.

Usher, M. B. 1986. Wildlife conservation evaluation. Chapman and Hall, London.

Vance-Borland, K., R. Noss, J. Strittholt, P. Frost, C. Carroll, and R. Nawa. 1995/96. A biodiversity conservation plan for the Klamath/Siskiyou region: A progress report on a case study for bioregional conservation. *Wild Earth* 5 (4): 52–59.

Vickery, P. D., M. L. Hunter, and S. M. Melvin. 1994. Effects of habitat area on the distribution of grassland birds in Maine. *Conservation Biology* 8:1087–1097.

Wagner, F. H. 1996. Ethics, science, and public policy. *BioScience* 46:765–766.

Wahlberg, N., A. Moilanen, and I. Hanski. 1996. Predicting the occurrence of endangered species in fragmented landscapes. *Science* 273:1536–1538.

Walter, H. S. 1990. Small viable population: The red-tailed hawk of Socorro Island. *Conservation Biology* 4:441–443.

Walters, C. J. 1986. *Adaptive Management of Renewable Resources*. McGraw-Hill, New York.

Walters, C. J., and C. S. Holling. 1990. Large-scale management experiments and learning by doing. *Ecology* 71:2060–2068.

Whitcomb, R. F., C. S. Robbins, J. F. Lynch, B. L. Whitcomb, K. Klimkiewicz, and D. Bystrak. 1981. Effects of forest fragmentation on avifauna of the eastern deciduous forest. Pp. 125–205 in R. L. Burgess and D. M. Sharpe, eds., *Forest Island Dynamics in Man-Dominated Landscapes*. Springer-Verlag, New York.

Wiens, J. A. 1986. Spatial scale and temporal variation in studies of shrubsteppe birds.

Pp. 154–172 in J. Diamond and T. J. Case, eds., *Community Ecology.* Harper and Row, New York.

———. 1996. Oil, seabirds, and science. *Bioscience* 46:587–597.

Wikramanayake, E. D., E. Dinerstein, J. G. Robinson, U. Karanth, A. Rabinowitz, D. Olson, T. Matthew, P. Hedao, M. Conner, G. Hemley, and D. Bolze. In press. An ecology-based approach to setting priorities for conservation of tigers, *Panthera tigris,* in the wild. *Conservation Biology.*

Wilcove, D. S. 1985. Nest predation in forest tracts and the decline of migratory song-birds. *Ecology* 66:1211–1214.

Wilcove, D. S., and D. D. Murphy. 1991. The spotted owl controversy and conserva-tion biology. *Conservation Biology* 5:261–262.

Wilcove, D. S., C. H. McLellan, and A. P. Dobson. 1986. Habitat fragmentation in the temperate zone. Pp. 237–256 in M. E. Soulé, ed., *Conservation biology: The sci-ence of scarcity and diversity.* Sinauer Associates, Sunderland, MA.

Wilcove, D. S., M. McMillan, and K. C. Winston. 1993. What exactly is an endangered species? An analysis of the U. S. Endangered Species list: 1985–1991. *Conserva-tion Biology* 7:87–93.

Wilcove, D. S., M. J. Bean, R. Bonnie, and M. McMillan. 1996. *Rebuilding the Ark: To-ward a More Effective Endangered Species Act for Private Land.* Environmental Defense Fund, Washington, DC.

Wilcox, B. A., and D. D. Murphy. 1985. Conservation strategy: The effects of frag-mentation on extinction. *American Naturalist* 125:879–887.

Williams, J. E., J. E. Johnson, D. A. Hendrickson, S. Contreras-Balderas, J. D. Williams, M. Navarro-Mendoza, D. E. McAllister, and J. E. Deacon. 1989. Fishes of North America endangered, threatened, or of special concern. *Fisheries* 14 (6): 2–20.

Wilson, E. O. 1985. The biological diversity crisis. *BioScience* 35:700–706.

———. 1992. *The Diversity of Life.* Belknap Press of Harvard University Press, Cam-bridge, MA.

Woodley, S. 1993. Monitoring and measuring ecosystem integrity in Canadian Na-tional Parks. Pp. 155–176 in S. Woodley, J. Kay, and G. Francis, eds., *Ecological In-tegrity and the Management of Ecosystems.* St. Lucie Press, Ottawa, Canada.

Woolfenden, G. E. 1974. Nesting and survival in a population of Florida scrub jays. *Living Bird* 12:25–49.

Woolfenden, G. E., and J. W. Fitzpatrick. 1984. *The Florida Scrub Jay: Demography of a Cooperative-Breeding Bird.* Princeton University Press, Princeton, NJ.

Wright, R. G., and D. J. Mattson. 1996. The origin and purpose of national parks and protected areas. Pp. 3–14 in R. G. Wright, ed., *National Parks and Protected Areas: Their Role in Environmental Protection.* Blackwell Science, Cambridge, MA.

Yaffee, S. L. 1991. Avoiding endangered species/development conflicts through inter-agency consultation. Pp. 86–97 in K. Kohm, ed., *Balancing on the Brink of Ex-tinction.* Island Press, Washington, DC.

INDEX

Adaptive management, 76–77, 117, 125, 207, 222
  assessing, in conservation plans, 133–36
  in Coachella Valley fringe-toed lizard HCP, 149
  in Coachella Valley MSHCP, 153
  in habitat-based conservation plans, 186–193, 197
  and "linear comprehensive management," 186–87, 188–89
  in NCCP, 141–142
  and "no surprises" policy, 50, 63–64
  and precautionary principle, 86
  use of indicators in, 194–96
  and viability of reserve systems, 69
Alligator, American, (*Alligator mississipiensis*), 39
ALEX (simulation program), 168
Animal rights movement, 20

Babbitt v. *Sweet Home Chapter of Communities for a Greater Oregon*, 12, 33
Balcones Canyonlands HCP, 41, 59, 71–72, 198
Bear, grizzly (*Ursus arctos horribilis*), 83, 103, 129, *130*
Beetle, elderberry long-horned (*Desmocerus californicus dimorphus*), 36
Big Cypress National Preserve, 185
Biodiversity, 1, 6, 39, 107-8, 207. *See also* Ecological integrity
Bioregions. *See* Landscape
Brevard County, Florida, HCP, 41–42, 58, 59, 142–49
  cost of alternatives under, 149
  role of scientists in, 54, 55
Bureau of Land Management, 150

California Department of Fish and Game, 150
Callipe silverspot (*Speyeria callipe callipe*), 36
Captive breeding, 37
Cleveland National Forest, 139
Coachella Valley, 37, 91, 149
Coachella Valley Association of Governments, 150
Coachella Valley fringe-toed lizard (*Uma inornata*), 36
  HCP for, 37, 59, 149, 183–84, 186
  monitoring plan for, 59, 133–134
Coachella Valley multiple-species HCP (MSHCP), 43, 149–53
  as equivalent to an NCCP, 150
Coastal sage scrub, 42, 43, *57*
  biogeography and inventory of, *60–61*
  Natural Community Conservation Plan (NCCP) for, 42, 70, 127, 139
    Central/Coastal Orange County subregion, *56*, 138–142
    implementation budget for, 59
    research agenda for, *60–62*
    scientific review panel for, 42, 81, 168–69
    survey, research, and conservation guidelines for, 42, 70–71
    See also Natural Community Conservation Planning program
  research questions, 164-66
Community, xv
Community conservation. *See* Ecosystem conservation
Condor, California (*Gymnogyps californianus*), 39
Conservation, scale of, 87, 91

Conservation biology, xii, 13, 73, 123, 162–63
 and advocacy, 78-80
 and biology, 91–92
 concern with target species, 92
 and environmental policy, 23
 as "mission-oriented" science, 75
 philosophical principles of, 74–76
 and "population thinking," 65–66
 principle of bigness, 92, 93–94
 and reserve design, 92–110, 177–183
 as science of case studies, 73, 155
 and scientific method, 157, 162–63, 166
 and uncertainty, 76–77, 80–81
Conservation legislation, 22
Conservation planning, 12–15
 agencies in, 223–24
 biotic and abiotic processes in, 184–86
 citizen activists in, 221–22
 developers and industry in, 222–23
 political obstacles to cooperation in, 198–99, 205–7, 213
 research questions for, 164–66
 role of scientists in, 52–57, 124–25, 211–212, 220–21
 scientific methodology of, 157, 162–63, 166
 See also Conservation plans; Endangered Species Act; Habitat-based conservation; Habitat-based conservation planning; habitat conservation
Conservation plans
 adaptive management capability of, 133–36
 attention to species or ecosystems, 127–28
 conservation of ecological processes, 131
 contribution to biological recovery, 112–15, 120–123
 contribution to ecosystem conservation, 126–33
 criteria for assessing, 111–154
 funding for science, 125
 goals of, 83, 87, 88–90, 180

 habitat in conservation management, 114–15
 habitat value increase, 113–14
 integration of science into planning, 124–25
 map-based analysis in, 126
 monitoring program of, 135
 ongoing threats, 122–23
 population size and viability, 115, 120–22
 research agenda and design, 126
 reserve design in, 123
 responsiveness to information, 136
 scientific rigor of, 123–126, 157, 162–63
 spatial scale of analysis, 131–32
 species coverage, 129–31
 temporal scale of analysis, 132–33
 testing management practice of, 135–36
 timely analysis in, 136
 umbrella species in, 128–29
Conservation and Recreation Lands (CARL) acquisition program (Florida), 143, 149, 205
Corridors, 4–5, 72, 102, 165, 199
 and reserve design, 61, 64, 72, 102, 181
Csuti, Blair, 39

Defenders of Wildlife, 9, 73, 200, 202
Department of Interior, 190. See also "No surprises" policy
Dolan v. City of Tigard, 35

Eagle, bald (Haliaeetus leucocephalus), 24, 39
Ecological integrity, 87, 91, 105
 indicators of, 152–53, 191–93, 194–96
 and species conservation, 206–7
Ecology, 22, 24, 73, 110, 157
 applied, 22–23, 162, 85
 development of, 3
Ecoregions, 200–4. See also Landscape
Ecosystem conservation, 38–43, 126–33
 alternative to species-by-species approach, 9, 39, 106–7

attractive to private sector, 40–41
consistent with intent of ESA, 42
principles for, 105–110
See also Habitat-based conservation
Ecosystem management, 6, 14, 23, 106,
    132, 176
and poorly defined goals, 15
and reserve strategy, 176
and uncertainty, 76
Ecosystems
complexity of, 2, 76–77
defined, xvi
determining boundaries of, 108–9
and ESA, 1
and listed and candidate species, 106–7
status of in United States, 9, 10
Edwards aquifer, 71
Egler, Frank, 76
Ehrlich, Paul, 75
"Eminent domain," 33
Endangered Species Act (ESA), xi, 1, 22,
    139, 214
conservation planning under, 12–13,
    14–15
critical habitat under, 12, 45
ecosystem conservation under, 1, 42,
    111, 222
exceptions to prohibition on harming
    listed species, 12
faults with, x, 1–2, 12–13, 137
general recovery goal of, 64
habitat conservation under, 23–25,
    32, 34–36
and habitat destruction, 1–2, 50
"harm," 11–12, 24, 33, 34
impact on national economy, 25
implementation of, 11–12, 23, 32, 34,
    45, 50
interpretation of recovery under, 67
listing species under, 6–8, 39, 89, 111,
    209
avoiding, 2, 29, 43, 150–51, 153,
    223–24
criteria for candidate species, 8, 98,
    111
elimination of category 2, 7–8
growth in list of candidate species, 7
listing of local populations, 93, 105

moratorium on, 7, 45
and poorly known species, 130–31
population sizes at time of listing,
    44, 45
prelisting agreements, 29, 218
See also Species, status of
1978 amendments to, 45
1982 amendments to, 25, 113
and non-federal activities, 25
and plants, 12, 24, 130–31, 137
and private lands, 13, 25, 26–32, 49,
    137–138, 214–219
and property rights, 33–35, 66–67
reauthorization of, 50, 58–59,
    214–219, 224
recovery plans, 11, 12, 32, 45–47, 65,
    67
and delisting species, 64
for California condor, 39
for desert tortoise, 66
for Florida panther, 83
goals of, 83, 87
for grizzly bear, 83
and HCPs, 32, 46–47, 68, 82,
    111–13
intent of, 67
lack of FWS funding, 46
legal status of, 67
on private property, 66–67
population goals of, 112
and "safe harbor" program, *30-32*
scientific quality of, 11, 46
single-species vs. multispecies, 11
recovery of various species under, 39
section 4, 42
section 7, 12, 24–25, 141
section 9, 12, 26, 138,
section 10, 12, 32, 64, 137, 143, 197
abuses of, 36
incidental taking under, 113
interpretations of Congressional
    intent in, 25, 32
not biologically conservative, 46
process analogous to Section 7 con-
    sultation, 25
proliferation of permits under, 59
single-landowner vs. regional per-
    mits under, 32

Endangered Species Act, section 10
(*continued*)
See also Habitat Conservation
Plans
shortcomings of standards, 65
"taking," 33, 34
defined, 11–12
incidental take permits, 12, 28
regulatory, 66–67
and species-habitat connection, 3
Endangerment
causes of, 7
criteria for. *See* IUCN Red List criteria for endangered species
Environmental Defense Fund, 30
Environmental education, 20, 90
Environmental impact statements, 156
Environmental Protection Agency
(EPA), 73
Everglades National Park, 185
Extinction, historic, 21

Falcon, peregrine (*Falco peregrinus*), 39
Ferret, black-footed (*Mustela nigripes*),
37–38
Fig wasp, 173–75
*First Evangelical Lutheran Church of Glen-
dale* v. *City of Los Angeles,* 33
Fish and Wildlife Service, (FWS), 6, 28,
199, 205
adopts "safe harbor" program, 30
approval of recovery plans, 46
and candidate species, 7–8
categories of HCPs, 82
Florida field office, 46, 146
handbook on HCPs, 29, 32, 46, 82
interpretation of "harm" under ESA,
11–12
investment in HCPs, 34–36
and NCCP, 42
and "no-take" agreements, 29
and Red Oak HCP, 36
review of Federal activities under section 7, 24–25
shift toward larger HCPs, 35
and status of species on private land, 8
treating HCPs as permitting process,
111

Florida Natural Areas Inventory, 88–89
Florida scrub HCP. *See* Brevard
Country, Florida, HCP
Forest Service, 150

Gap analysis, 43
Gap Analysis Program, 11
Geographic Information Systems (GIS),
60–61, 107, 123, 171, 193
in Brevard County HCP, 147–148
in Coachella Valley MSHCP, 152
as conservation tool, 160–62
in NCCP, 60–61, 62
and reserve design, 157
Georgia Pacific Corporation, 30
Glen Canyon Dam, 185
Gnatcatcher, California (*Polioptila californica californica*), 43, 61, 127,
139, 165, 205

Habitat
area requirements for various species,
4–5
blocks of, 99–103
defined, xv
as proper focus for conservation, 2–5
quality vs. quantity, 114
Habitat alteration, 5–9
as form of "harm" to listed species, 12
human capacity for, 19
lack of public understanding of, 19–20
limited legal constraints to, 1–2
as primary cause of extinction in
North America, 21
Habitat-based conservation, 9, 11–12, 14
reconciling with species conservation,
15, 16–17
scientific principles of, 73–110
See also Conservation plans; Endangered Species Act; Habitat-based
conservation planning; Habitat-
based conservation plans;
Habitat conservation
Habitat-based conservation planning, xi,
12–13, 14–15, 23, 209–210
and recovery, 45
linking disparate planning efforts, 197
preventing listings under ESA, 221–22

See also Conservation plans; Endan-
    gered Species Act; Habitat-based
    conservation; Habitat-based con-
    servation plans; Habitat conser-
    vation;
Habitat-based conservation plans
    as case studies, 73–74, 207
    and conflicts between conservation
        and economic development, 50
    components of, 155–208
        assessment of population status
            and viability, 166–76
        reserve selection and design,
            176–83
        scientific methods, 156–57,
            162–63, 166
        sustenance of ecological and evolu-
            tionary processes, 183–86
        unifying disparate planning efforts,
            197–99, 205–7
    criticism of science in, 49–72
    defined, xv
    and growth management, 210–11
    and recovery plans, 112–13, 153
    See also Conservation plans; Conser-
        vation planning; Endangered
        Species Act; habitat-based con-
        servation; habitat-based
        conservation planning; habitat
        conservation
Habitat connectivity, 102, 140–41
    in coastal sage scrub, 165
    measures of, 193
Habitat conservation
    under ESA, 1–2, 23–47
        first major report on, ix, xi–xii
    history of, early, 21–23
    incipient movement towards, 20
    as proactive strategy, 1–2
    scientific foundation of, xi
Habitat Conservation Plans (HCPs), 12,
    14, 25
    and antienvironmental sentiment,
        143, 148
    Conference Committee Report on
        (1982), 32
    FWS categories of, 82
    FWS handbook on, 29, 32, 46, 82

and hotspots, 108
inadequate mitigation requirements
    for, 32, 36
incidental take permits issued, 34
and lack of data on species of concern,
    70
and large-scale biological uncertainty,
    71–72
limitations of, 32, 37, 41, 137–38
mitigation requirements under,
    49–50, 59, 63–64
multispecies HCPs, 35, 38–43, 40–41,
    216–17
    vs. natural community or eco-
        system level plans, 41–42
    as undermining ESA, 64
number in development, 34
peer review of, 56, 219
as permits to destroy habitat, 111
and recovery plans, 32, 46–47, 66–68,
    111–13
and reserve systems, 68–69
scale of, 34–36
scientific input and review, 51, 52–56,
    56–59, 219
single-species HCPs, 36–38
small, 34–36, *218*
    biological inadequacy of, 58–59
    scientific scrutiny of, 81–82, 124
specific plans
    Balcones Canyonlands HCP, 41, 59,
        71–72, 198
    Brevard County, Florida, HCP,
        41–42, *54*, *55*, 58, 59, 142–49
    Coachella Valley fringe-toed lizard
        HCP, 37, 59, 183–84, 186
    Coachella Valley MSHCP, 43,
        149–53
    Red Oak HCP, 36
    Stephens' kangaroo rat HCP, 40
standards for approval (jeopardy
    standard), 25, 32, 46, 50, 64, 113,
    137
and uncertainty, 63–64, 68–69
WWF report on, xi
Habitat fragmentation, 67, 93, 99–102,
    165
Habitat suitability indices, 114

Habitat types, xv
Hypothesis testing, 156–57, *158–59*, 162, 164–66

Indicator species, 39, 54, 106, 117, 192
Island biogeography, 92, 99–100
IUCN Red List criteria for endangered species, 98, 108, 173, 174–75

Kangaroo rat, giant (*Dipodomys ingens*), 206
Kangaroo rat, Stephens' (*Dipodomys stephensi*), 36, 40, 43
Keystone Bituminous Coal Association v. De Benedictis, 33
Keystone species, 15, 119, 121

Lacey Act, 22
Landscape, xvi
    linkages. *See* Corridors
Land trusts, 20
"Law of the minimum" (Liebig), 3
"Law of tolerance" (Shelford), 3
Leopold, Aldo, 20, 24, 75
Liebig, Justus von, 3
Lucas v. South Carolina Coastal Commission, 34

Melitaea cinxia, 171
Melitaea diamina, 171
Merritt Island National Wildlife Refuge, 143
Metapopulation, 132, 167, 168, 171
    of California gnatcatcher, 168
    defined, xvii
    theory of, 100–1
Migratory Bird Treaty, 22
Minimum viable population, 94, 96–99
Mission blue butterfly (*Plebejus icarioides missionensis*), 36
Monitoring, 131, 133–35, 194–96
    ecological, *90*, 186–87, 190–93, 197
    in small HCPs, 59
    in NCCP, 60–62
    population, 37–38, 67
Multispecies conservation, 40–41, 127, 132

Murrelet, marbled (*Brachyramphus marmoratus*), 102, 128

National Environmental Policy Act (NEPA), 22, 135, 156
National Forest Management Act (NFMA), 22–23
National Marine Fisheries Service (NMFS), 6, 24
National Park Organic Act (1916), 22
National Park Service, 150, 205
Natural community, xvi
Natural Community Conservation Planning Act, 138
Natural Community Conservation Planning program (NCCP), 41, 42–43, 123, 138, 197, 205
    conservation guidelines for, 56, 58, 140
    as means of avoiding listing of species under ESA, 43
Natural community conservation plans (NCCPs), 13
    and reserve systems, 68–69
    standard for approval (jeopardy standard), 6
    as undermining ESA, 64–66
Nature Conservancy, The, xvi, 55, 56, 88, 89, 210
    and "fine filter/coarse filter" approach 39–40, *116*
    and species status, 8, 89, 201
New Forestry, 176
Niche, ecological, 3
Nollan v. California Coastal Commission, 34
"No surprises" policy, 16, 28, 40, 50, 63, 215–16
    and adaptive management, 134–35
    preventing negative impacts from, 190
"No-take" agreements, 29

Panther, Florida, 83
Peer review, 53, 56, 74, 75, 166
Pelican, brown (*Pelecanus occidentalis*), 39
Physical habitat, xv

Pine, longleaf, 30–31, 131
Population
    defined, xvi
    effective size of (N_e), 97
    peripheral, 104–5, 107, 171
    source-sink, xvi, 4, 30–31
    viable, xvi–xvii, 3–4
Population viability analysis (PVA), 67,
        115, 120, 122, 176, 220
    alternatives to, 115, 170–73
    and mutualism, 173–75
    of northern spotted owls, 167
    problems with, 168–69, 170
    spatially explicit, 132, 155–56,
        166–70, 178
    of tigers, 173
    use of criticized, 39
Precautionary principle, 81, 84–86
Private land, incentives to conservation
        on, 26–32
    regulatory assurances regarding, 16,
        40, 63, 138, 143, 151–52, 222
    See also Endangered Species Act; "No
        surprises" policy
Prototyping, 207

Recovery plans. See Endangered Species
        Act
Red Oak HCP, 36
Regions. See Landscape
Representation, 11, 43, 83
Reserve selection and design, 13, 23,
        57–58
    and adaptive management, 64
    and assessing conservation plans, 123
    in Brevard County HCP, 147–48
    buffer zones, 109–10, 177, 181
    in Coachella Valley MSHCP, 151
    and global warming, 71–72
    and GIS, 157
    guidelines for, 92–107
    in habitat-based conservation plans,
        176–183
    habitat blocks in, 99–103
    human visitation, 182–83
    and hypothesis testing, 158–59,
        176–177
    implementing, 178–183

    in NCCP, 138–141, 168–169
    and northern spotted owl, 158–59
    principles of, 93–94, 96–105
    and target species, 115, 120–122
    and uncertainty, 69
    zoning in, 90, 109–110, 176
Reserve networks, 4–5, 12, 20, 181, 182
Reserves, forest, 21, 22
Restoration, 99, 180, 181, 185, 186, 191,
        206

"Safe harbor" program, 30–31, 217
San Bruno Mountain, 25, 36, 137
Scott, J. Michael, 39
Scrub, Florida, 41, 145, 171, 173
    species in, 38, 145
Scrub jay, Florida (Aphelocoma
        coerulescens), 35, 46, 132
    data on, 127, 146
    as umbrella species, 146
    see also Brevard County, Florida,
        HCP
Scrub, rosemary (Ceratiola ericoides),
        173
Shelford, Victor, 3
SLOSS (single large or several small),
        92
Soulé, Michael, 75
Species
    area requirements of, 4–5, 94–95, 121
    flagship, 120
    game, 22, 92
    indicator, 39, 54, 106, 117, 191–192
    keystone, 15, 119, 121
    population variability of, 103–4
    status of, 8, 9
        IUCN Red List criteria, 98, 108,
            173, 174–75
        on private land, 8, 13, 214–19
    target, 54, 91, 116–121, 127–128, 192
        and conservation biology, 92
        dispersal behavior of, 99, 100
        and NCCP, 61, 142
        persistence of 115, 120–122
        and rescue effect, 100
    types emphasized in conservation
        planning, 116–21
    umbrella, 106, 128–29, 130, 146

Species conservation, 16–17, 95–97, 92–107
  single-species conservation plans, 36–38
    limitations of, 14, 39, 126–31
    as only short-term option in some cases, 37
  threats to small populations, 95–97
Spotted owl, northern (*Strix occidentalis caurina*), 27, 58, 102, 128, 193
  Interagency Scientific Committee for Conservation of, 8, 93, 103, 156–57
    reserve design principles of, 93–94, 96–102, 158–59
  PVA of, 167

Target species, 54, 91, 116–121, 127–128, 192
  and conservation biology, 92
  dispersal behavior of, 99, 100
  and NCCP, 61, 142
  persistence of 115, 120–122
  and rescue effect, 100

Thoreau, Henry David, ix, x
Thomas, Jack Ward, 158
Tortoise, desert (*Gopherus agassizii*), 66
Translocation, 37

Umbrella species, 106, 128–29, 130, 146

Vireo, black-capped (*Vireo atricapilla*), 111

Warbler, golden-cheeked (*Dendroica chrysoparia*), 198
Wildlands Program, The, 210
Wilson, Edward O., 75
World Wildlife Fund, ix, xi, 200, 202–3
Wolf, Mexican gray (*Canis lupus*), recovery plan for, 46
Woodpecker, red cockaded (*Picoides borealis*), 29–31, 36, 46, 172
Wren, winter (*Troglodytes troglodytes*), 102

Yellowstone Park Protection Act (1872), 22